A Guide to... Asiatic Parrots in Australia

By Sid & Jack Smith

Published and Edited by Australian Birdkeeper ©

First Published 1990 by
Australian Birdkeeper P/L
PO Box 6288,
South Tweed Heads,
NSW. 2486. Australia.

ISBN 0 9587455 2 8

Front Cover:
Top left: Cock Derbyan
Btm Left: Cock Lutino Ringneck
Btm Centre: Grey Indian Ringneck
Btm Right: Cock Alexandrine

Design, Type and Art: Jackson Studios (Gold Coast)
Colour Separations: Haighs Foto Art (Brisbane)
Printing: Prestige Litho (Brisbane)

Contents

Acknowledgements

Special thanks to Dr Rob Marshall for his section on Disease Prevention and Cures.

Photographic Credits

All photographs by **Wilson Moore** except for:

Jack Smith:	Page 18 - Top left
	Page 22 - Bottom
	Page 38 - All photos
	Page 42 - Bottom
	Page 48 - Top left, Bottom left
Peter Odekerken:	Page 44
	Page 45
Kevin Wilson:	Page 22 - Top left
	Page 23 - Top left, Bottom right
Tod Osborne:	Page 39
	Page 40 - Centre left
Ian Brown:	Page 40 - Bottom left
Henry Koesling:	Page 34 - Bottom left

Asiatic Parrots

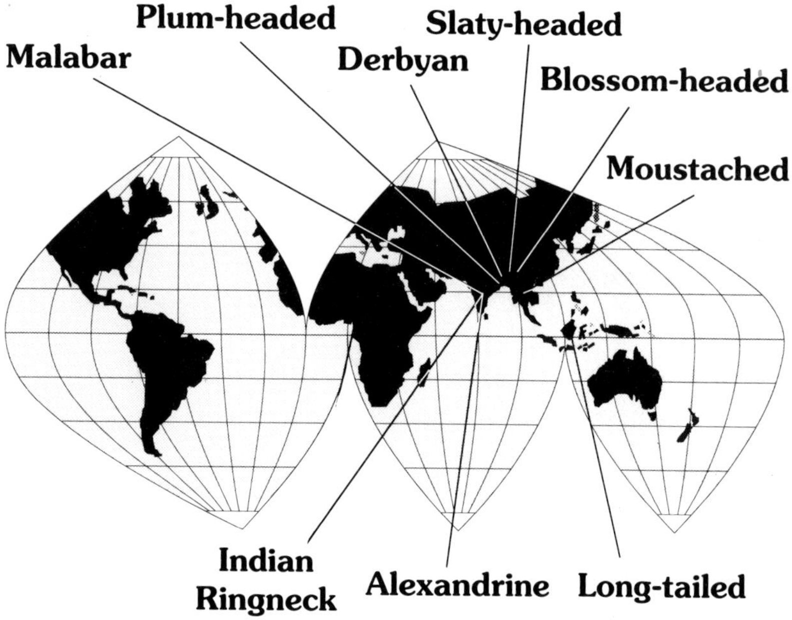

From the Authors

Aviculture is a wide and fascinating hobby that draws all types of people, and over the years we have formed some wonderful friendships. Our wives, Gwen and Glad must be thanked as we all have our particular jobs to attend to.

Our hobby could not have expanded to what it is today without each others help.

We have also gained a lot of knowledge by meeting and visiting other fanciers past and present, exchanging ideas and methods and must thank them all for their hospitality.

Syd & Jack Smith

An Introduction to Asiatic Parrots

Introduction

Our background to aviculture started in our early days when we would spend weekends roaming the countryside either rabbiting or bird nesting.

Hours would be spent sitting looking and listening in the scrub or around water-holes at bird life and no doubt we both gained a lot of knowledge on bird behaviour.

Returning home when we got hungry, with perhaps young birds, rabbits or snakes, our mother would give us both a good dressing down. She never did get through to us that weekends were for cutting firewood and lawn mowing. After accumulating birds of all descriptions housed in a conglomeration of netting and fruit cases, our father built our first aviary, a 1.20m (4ft) square box with a wire front on legs 60cm (2ft) off the ground and we are still building today.

Now, having over 200 breeding flights at Harkaway which is 27 miles east of Melbourne, Victoria, we are both kept busy after work and at weekends attending to our hobby.

Having kept most parrots that have been available in Australia, except for the Lorikeets and Black Cockatoos, you soon find some require more attention than others, so it is a matter of what time you have available and what species fit into that schedule.

We know most bird people would like to keep all types of birds but there is not much point in keeping something if you can't give it enough attention to be successful with it.

Asiatics fit into to our time programme and that is why our collection houses so many. Basically they are hardy, will breed on an easily prepared diet, have a very tight feather pattern which gives them an immaculate appearance, and all breed at the one time of the year.

Over the years that we have had Asiatics there has always been a steady demand from other aviculturists as nobody can mass produce them by trapping which can happen with native species.

Management

Acquiring Stock

In acquiring birds we always like to start with young stock obtained from other breeders whose climate and housing is similar to ours. With young birds, you may have the problem of not knowing what sex they are but at least you know they are not somebody elses culls and have a full life ahead of them. If you obtain birds from perhaps a warmer climate or they may have been bred in closed-in heated or box type aviaries, you have the hassle of acclimatising them to your conditions.

All too often, one hears of people obtaining adult breeding pairs even in the breeding season and in most cases they are only worn out old culls. Certainly genuine adult birds do come on the market at times due to all sorts of circumstances but it is advisable to go well into their background before purchasing such birds.

A good time to purchase stock in Australia is just after Christmas. The breeding season has finished and there is a good choice of young birds on the market. The days are still long and the weather is warm. This gives new arrivals time to settle in and you usually have more time to observe their behaviour in their new aviaries.

A trap many aviculturists fall into is that they don't keep any replacement stock from their own breeding. Fortunately, having plenty of aviary space, we always put a few youngsters aside for our own use, one never knows when you may lose a breeder and this is when you find out how hard some species are to replace - after spending hours on the phone and countless miles trying to find the right bird.

A good carry box is a must for all aviculturists. It is surprising the number of people who have the money to buy expensive birds yet put no thought into boxes for transportation. The box we use is made of 18mm (3/8in) ply board, 38cm (15in) long, 25cm (10in) deep and 23cm (9in) high with a weld mesh front. A door big enough to put your hand in with a good type latch that doesn't jiggle open on long journeys, saving the embarrassment of the birds flying off over the horizon.

With large birds, one per box is a good policy, as with two in a confined space fighting can take place resulting in all sorts of injury. This is especially the case with Alexandrines.

For dispatching stock we use clean ply board boxes built to suit the species - reasonably low with limited viewing area but enough to allow air flow. We prefer to nail the top down with no doors to be opened. Address it clearly with name, phone number etc. Also complete sender's address. If permits are required, these should be attached to the box. Always check with the airlines regarding flight times and advise receiver at least 24 hours in advance. We have dispatched many consignments within Australia and overseas with no problems if these few guidelines are adhered to.

Housing

Building aviaries is something we have done quite a bit of and after extensive travelling and looking at all types of constructions, we still prefer the open flight type aviary attached to a solid shelter.

All areas are different with regards to location but we prefer aviaries to face east as the occupants get the morning sun and this gets them active for the day ahead. We believe the sun is very important in stimulating birds to breed along with having access to natural rain to maintain feather condition and provide moisture for incubation. Remember Asiatics don't bathe in the water bowl like many native parrots do.

In our area we experience a lot of cold damp days with no sun for days on end so we have found it necessary to build the frame of the shelter with hard wood timber and the divisions of cement AC sheeting. The ceiling is also lined with AC sheeting, helping to insulate against condensation directly off the corrugated iron roof.

We find a shelter shed of this type is far warmer and maintains a more even temperature than the commercial type aviaries that are manufactured to a price, out of flat iron and pop rivets. They can become either freezing cold or excessively hot.

For the flight section 20mm (3/4in) galvanised pipe is used for the frame as this is far stronger than 12mm (1/2in) and one is able to strain the netting on neat and tidy without bowing the frames. Maybe it is because we are used to it, but we prefer to tie-wire netting onto galvanised pipe rather than to pop rivet weld mesh onto box steel.

I know it is getting harder to obtain good quality netting nowadays and the next alternative is to use weld mesh. However, I have found that birds damage their feathers on weld mesh more than they do on netting. For Asiatics the size of weld mesh can vary. The choice should be made with a view to keeping undesirable creatures out (including wild finches) rather than the Asiatics in, so always buy as small as you can afford. Minimum gauge or thickness should be 17 gauge or 1.3mm.

A safety passage is necessary on a breeding block of aviaries and we prefer one at the rear, wide enough for a wheelbarrow to be used for cleaning out etc. Full height doors also save the back and a lot of head butting. Birds fly forward out of the way when attending to feeding from the rear.

Roofing one third of the flight section adjacent to the shelter with fibre glass sheeting provides more protection and is an ideal area to hang nest logs or boxes.

In building aviaries, vermin proofing should always be considered and the less places for rodents to get established the better. The floors of our shelters are of concrete and the flights earth. If we find evidence of rats or mice we use poison in bait boxes along with regular digging and hosing of their tunnels. You just have to be as persistent as they are in order to keep their numbers down. As well as being disease carriers mice attract cats and snakes which don't mix with birds either.

Suspended cages are very fashionable nowadays, being easy to clean, cheaper to construct and many more breeding pairs can be kept in a small area. However, in the short time they have been in use, I have already had reports from America that a lot of larger type parrots (African Greys, Major Mitchells etc) reared under these conditions are very poor flyers and feather problems are on the increase. Although Ringnecks can be bred in such cages 60-90cm (2-3ft) wide, 1.80m (6-9ft) long and 1.20m (4ft) high, I would suggest they be housed in larger, more conventional aviaries in the non-breeding season so the birds can be kept in peak condition.

A row of Asiatic flights. The galvanised pipe delivering water can be seen running inside along the front.

▲ *A bank of flights.*

◀ *A view into the shelter.*

▼ *The rear walkway. Flights are on the left.*

◀ *Inspection doors are a must with heavy logs. Young Alexandrines can be seen here.*

Water is provided by running a 20mm (3/4in) galvanised pipe the full length of the breeding complex with a 45mm (3/16in) hole drilled adjacent to each water bowl. It is then just a matter of turning on the supply tap and water spurts out into each bowl. Bowls still have to be regularly scrubbed out though.

To gain maximum flight, perches are provided, one in the shelter and one at the end of the flight. We use natural black tea tree branches which when dry are extremely hard and last quite a while before they need replacing.

Cleaning in and around seed dishes takes place at least every week but we have a major clean up after the moult around March, when the shelters are cleaned of all seed and feathers, the walls scrubbed and the dirt floors of the flights raked over. At this time, liberal amounts of lime are spread around and dug in, helping to sterilise the floor. Birds remain in their flights while doing this.

Feeding

We will probably be criticised by many breeders for not feeding a better variety of food stuffs than we do. However, we just do not have the time to give all sprouted seed, fruit, vegetables, vitamin supplements etc in the variety and amounts some breeders advocate.

If we were to feed all these wonderful things to the amount of birds we have, it would result in all day being taken up with preparing and feeding.

We know of birds that have responded well to a smorgasbord type diet and we know of birds that have deteriorated on such a diet. The majority of reports of feeding a rich, varied diet to pairs of birds have shown no changes for the better to either the pairs or to the number or physique of their offspring. So providing your birds are receiving an adequate diet we feel anything more is of little benefit.

When feeding essential extras, I think it is important that whatever is used must be fed on a daily basis. Where many problems arise is when a certain fruit or vegetable is not available or may be of a poorer quality. This can easily upset your birds. Feed what you have time to prepare and use foods that can be obtained most of the year. If we believed our birds were suffering in any way from a lack of something in their diet then we would take steps to remedy the situation. Many of our birds have been breeding successfully with us since the 1960's.

Working in an apple orchard, Syd often observes the feeding habits of wild Rosellas. No attempt will be made to attack the crop until the fruit is fully ripe, when the sugar content is high. Nowadays, for cold storage purposes, we pick fruit before it reaches maturity hence no wild bird problems. The point we are trying to make is that fruit out of cold storage wouldn't be eaten if it was still on the tree, therefore apart from about eight weeks of the year it would be virtually impossible to obtain apples to the birds liking. Fruit seems to go off quickly and this uneaten fruit must be removed to avoid bacteria problems. Sprouted seed falls into the same category. One infected batch and you may lose all your youngsters for the year. We are not suggesting you should not feed sprouted seed to your birds, it is just that under our current circumstances we do not have the time to prepare these foods properly and so to avoid possible risks to our birds' health we do not include them in our diet.

Our diet consists of a basic seed mix of 2 parts grey sunflower, 1 part oats and 1 part budgie mix. The budgie mix is made up of 50% plain canary and the other 50% pannicum, Japanese and white millets. This is fed once a week, either in self feeders or open dishes on the floor twelve months of the year. I am of the belief that birds put themselves on a diet, as at certain times of the year it is quite noticeable that some seeds are preferred more than others.

In addition to seed, daily supplies of corn on the cob and milk thistles and oats in head are fed especially once youngsters start hatching, this being continued four to six weeks after the young fledge. We find at this stage, parents and young lose interest in the corn so the quantity fed is very much reduced until the next breeding season.

Corn cobs are brought in fresh and are frozen until needed when the amount required is thawed out the night before. Placing the cobs on 7cm (3in) nails driven into the perch, they are completely devoured during the day and provide an excellent source of bulk food that parents can regurgitate to their young. If frozen cobs are used up before the end of the breeding season they can be substituted by boiling corn grain for two hours. This is then rinsed and a handful given to each breeding pair. This corn grain must be boiled until soft as compaction can occur in youngsters' crops under ten days of age. Shell grit and cuttle fish bone is also provided as necessary. This simple diet is the same each day and we are very satisfied with the results we have achieved. This is our basic diet, any other food stuffs such as fruit or seeding grasses are given on a casual basis - more as treats than anything else.

Breeding

To be successful in breeding Asiatics, or in fact any form of livestock, one must keep accurate records and have a strict culling programme. If you don't, you will soon find that after a few years your stock will deteriorate. Breeding performances will be poor and it is not long before your collection is just a lot of has beens.

If breeding mutations, young should be closed rung between eight and fifteen days of age for identification purposes and records kept of parentage. If this is done, stock can be sold and guaranteed to the purchaser with confidence. If not closed rung, all sorts of things happen from birds being switched, doors left open and becoming mixed up and so the list and excuses go on.

Metal split rings that are placed on adult birds often cause problems as the bird, not being used to having a ring on, naturally tries to get it off and in doing so squeezes it tighter often cutting off circulation and the leg can be lost. There are split rings available made of stainless steel that are applied with a special pair of pliers and can't be squeezed tighter by the bird. Unless it is absolutely necessary, be careful about using metal split rings.

Over a period of years, you can improve your stock's type and colour by selecting the better birds to breed with. Some people never know a good bird from a bad one. We are always on the look out for better type birds in other collections and are ready to compliment their owners. You must never think you have the best.

In building up a collection we find that your chances of success are increased if you have at least three or more breeding pairs of a species. It is then possible to compare each breeding performance and you are able to select young for replacements from pairs that lay a good size clutch, have a high fertility rate, have good mothering ability and raise young that are true to type with no deformities as most of these things are hereditary. It is up to the breeder to take advantage of these characteristics.

Culling should be applied to birds that are poor parents and have small clutches, poor fertility, weak eye sight, crossed wings, turned in feet and weak bodied birds that sit low on the perch and so on.

Odd coloured feathers are usually the first signs of weakness or deficiencies and should also be avoided. Stock bred out of season is another no no. They are often weaker. Their parents should have bred during the season proper so out of season birds are often from third clutches. Many are hand-reared or reared by foster parents because few Asiatics regularly double or triple brood. What has to be remembered is that they are hatching from the ninth to fifteenth egg laid by the hen in the one year. Such eggs are usually never as good 'food wise' as the first eggs laid. So these chicks are always off to a bad start in life and usually don't improve.

Always remember your stock can only respond to the treatment they are given and poor performances are often the fault of the owners.

Pair bonding is another important part in the successful breeding of any bird. It may take a number of years before some types attempt to nest but with most Asiatics, they pair up reasonably easily.

▼ *A planted aviary is suitable for Plum-heads or Slatey-heads.*

▲ *Iron sheeting above and below ground deters rodents.*

▼ *Millet sprays are allowed to grow in these Ringneck aviaries.*

▼ *Asiatics breed well in aviaries like these.*

In breeding Indian Ringneck mutations, we have on many occasions broken up pairs for a particular mating with no problems. They settle down with their new partner within weeks and breed successfully.

In our area the Asiatics mainly breed between July and October.

Courtship and mating usually starts three to four weeks before egg laying and if you haven't seen Asiatics displaying you could be excused for thinking they were having an epileptic fit. With eyes rolling and dilating and the body stretching and bowing while taking short hops on the perch - their performance is quite different to our native species.

Mating usually takes place soon after this display and we have observed them mating on the ground and on top of the nest log but normally on the perch, not bothering them whether it is swinging or fixed.

Mating occurs several times a day from dawn til dusk, often giving a performance in front of visitors. Syd's young son explains it rather well with the words "I think they like doing that, Dad".

Approximate incubation times are given under the individual bird section in this book as it is hard to do a thorough check because although the hen is in the log it doesn't mean she is actually incubating the eggs. After ten days to a fortnight, inspection should be made to see if eggs are fertile. If not discard them and there is a chance a second clutch will be laid. If eggs are removed for fostering out, incubation or because they are clear, there is a good chance Asiatics will lay a second clutch. However, once the parents are feeding young there is much less chance of a second clutch being laid even if the young are removed for hand-rearing.

Nesting Logs or Boxes

Nest logs are now used in all our aviaries with the larger type parrots, who love to chew, because they last a lot longer. In our early days we used nest boxes of sawn timber or ply board and one was forever replacing them. There is nothing worse than the bottom falling out or a hole appearing in the side of the box due to the hen chewing away from the inside. This usually happens the day young hatch and it is pouring with rain resulting in young and the breeding season being lost for another twelve months.

We spend many hours of a night in the shed preparing hollows for the coming breeding season. Making use of natural knot holes and cutting them to various lengths to suit different species. We get a lot of satisfaction out of the end result and we are sure a lot of our breeding successes are attributed to these natural hollows.

A 38cm (15in) strap of 2.5cm (1"in) flat iron is bolted to each log so that it can be hung up with no risk of falling.

Arguments still arise on how logs or boxes should be hung, either vertically or on an angle. All our logs are hung vertically with very few problems. If you have hens that are clumsy in the nest, breaking or cracking eggs due to entry, we have found they are not worth persevering with as they often don't make good mothers and the offspring are likely to be the same. It is far simpler just to cull the hen and replace the bird with one that doesn't do it.

Of all the types of material available for use in a nesting chamber, we still prefer rotted hardwood sawdust. A supply of this is obtained from a saw mill and left in the open to decay. We find it holds a certain amount of moisture and can be rammed down firmly in the log, not becoming too dusty.

Nest logs are left in the aviary all the year round as Asiatics often use them if hawks and the like disturb them.

Sawdust is renewed soon after the shortest day (21st June) in preparation for the coming breeding season and regular checks should be made up until eggs are laid to see that the sawdust hasn't all been carted out or stones and other hard objects haven't been placed in the log by the birds.

Sickness

No matter how small your involvement in bird keeping, attempts should be made to broaden your knowledge on how to detect a sick bird and what can be done. We make a point of checking each aviary every day, if this is done, nothing seems to go wrong.

In twenty four hours a bird can deteriorate very quickly and if you only check your birds of a weekend, it is often too late. Birds can suffer from stress - perhaps just being moved from one aviary to another, night frights caused by cats or owls, bright car lights, sudden changes in temperature to name just a few.

If a bird is off colour they usually fluff up and generally look listless. If showing these signs and sitting on two legs they are not well. Birds perching on one leg are usually just resting.

In our safety corridors, a small hole is drilled through the wall so that you can observe the birds without disturbing them. This way you will soon decide if they are off colour or not. It is important to remember that if you find a sick bird, the quicker you take action the better your chances of its recovery.

Refer to the chapter on Diseases for more detailed information.

Conclusion

For the amount of time Asiatics have been in this country, many people wonder why they haven't become more freely available. We believe it is due to the fact that very few aviculturists have been seriously keeping and breeding them for any real length of time. It takes a number of years to build up pairs and many bird keepers do not start serious bird breeding until around retiring age and often, unfortunately, run out of time (not only birds drop off the perch!) resulting in the collection being split up and disposed of.

Nowadays people seem to specialise more in a particular species thus increasing the numbers and making a worthwhile contribution to aviculture. We are very fortunate to have the amount of Asiatics in Australia that we do and all efforts should be made to ensure that they don't die out.

▲ Green, Blue, Lutino and Albino babies.

Cock Blue Ringneck. ▶

Pair of Lutinos. ▶

◀ Cock Lutino.

◀ Albino cock.

▼ Albino hen.

Species of Asiatic Parrots

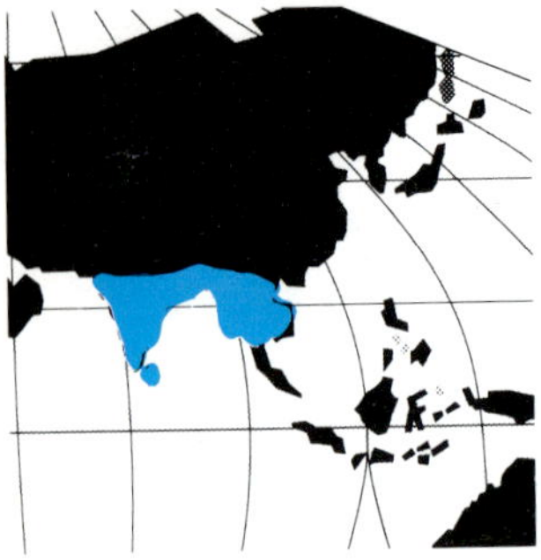

Alexandrine Parrot

Psittacula eupatria

The Alexandrine Parrot is certainly one of the largest of the Asiatics in Australian aviaries and because of its size, is not always suitable for every aviculturist. If you have a small back yard in suburban areas, they can be rather noisy at times causing problems with neighbours that don't like loud piercing bird calls.

However, a big percentage of aviculturists that live in open areas where noise is not a problem find they are an excellent bird. In fact, it has only been in recent years that people have taken notice of them, prior to that we feel they were very under-rated.

Like all Asiatics, rarely do you see one in poor feather condition apart from when they are moulting. Alexandrines are an excellent example of a good feathered bird. Often at bird shows the common remark from Budgerigar breeders is "how do you get a bird that size in perfect feather?" Our answer to that is, if they are housed properly, it just seems to be in their make up.

People new to aviculture probably don't realise how lucky we are to have these types of birds available. Thirty years ago there were very few Alexandrines about, but with careful breeding from dedicated aviculturists Alexandrines have slowly progressed and are freely available today. We think the point should be made that there is rather a marked difference in the size of some Alexandrines you will see to others. People will comment that there is a large race of the species and a small version. However, in our experience we would simply describe the difference in size as being the weak and the strong.

Alexandrine Parrots are generally excellent parents, rearing their young without problems - provided that the parents have enough food available before them to feed the youngsters. If they haven't, rarely will they let the young die as a lot of other birds would, but continue to rear them as best they can. On leaving the nest, the young are much smaller than they should be.

A good sized cock Alexandrine should measure around 55-58cm (22-23in) in length and be rather bold in the head and stand firmly on the perch with an alert appearance. Hens are slightly shorter in the tail and have a rounder, more feminine head than the cock.

In selecting birds, particular attention should be made to size and the amount of chestnut red on the wing patch of both sexes. Some Alexandrines are very poorly coloured in this area.

All our aviaries are of the open flight type as we are very firm believers that birds must have access to rain. Regardless of the time of year, as soon as it is raining, Asiatics delight in hanging on the wire or sitting on the perch to get a good soaking. We believe this is very beneficial, not only for maintaining their feather condition but also it is a great help in egg hatchability.

Alexandrines are good at chewing so an aviary built of galvanised pipe or steel tube with heavy weld mesh is needed at least 1.6mm or 16 gauge. The length should be at least 4.80m (16ft) - one third of this shelter) by 1.20m (4ft) wide and 2.10m (7ft) high.

We have seen a number of pairs housed together and they have bred successfully, but we only house a single pair per flight.

Natural hollow logs hung vertically are used for nesting. These being 90cm (3ft) long with an inside diameter of 25-30cm (10-12in). A 12cm (4.5in) entrance hole is cut 15cm (6in) from the top of the hollow. Logs of this size are rather heavy so an inspection opening is made 23cm (9in) from the bottom. This allows the breeder to inspect without having to lift the log down. Natural rotted hardwood sawdust is used for the nesting chamber.

If logs are not available, a box made of sawn timber or ply board covered with flat iron to restrict the birds from chewing with a ladder of weld mesh or similar attached inside would be suitable. It should measure approximately 25cm (10in) by 25cm (10in) by 90cm (3ft) high.

Most young Alexandrine cocks will show colour at twelve months of age yet others don't colour until eighteen months but usually have a full neck ring of pink and black by the time they are two years old. Hens don't get a neck ring, only the wing patch.

We never expect breeding success until both cock and hen are at least three years old. From this age on we have found them to be very reliable breeders usually going to nest in Melbourne around the end of July. Three to four eggs would be a normal clutch and although cocks will enter the log, only the hens incubate.

Incubation lasts about twenty four days and on hatching young soon need a constant supply of food, which can be easily regurgitated by the parents. A basic diet of grey sunflower, oats and budgie mix is always available along with a daily supply of green food such as milk thistle or oats in head which we find are excellent, and corn on the cob or boiled corn, whichever you are able to obtain. I have found that if a daily supply of corn is not supplied, the parents are not able to keep up the food required just out of a dry seed mix. I don't feed a lot of fruit but apples will readily be eaten.

As mentioned before Alexandrines seem to be excellent feeders and usually rear every young one hatched. Babies stay in the nest for about seven to eight weeks.

Being such large birds and heavy fliers, it is advisable to place a leafy branch or something similar at either end of the flight that youngsters can crash land into, saving skinned heads or broken wings. Within a few days they soon settle down and are a sensible bird in a long flight.

Young are independent a month after leaving the nest but we have never had to remove young from the parents' flight through bullying as it is a rare occasion for Alexandrines to double brood.

In an effort to produce Lutino Alexandrines, some breeders have crossed them with Lutino Indian Ringnecks, however it is a long programme and care must be taken to keep thorough records of matings and hybrids. The end result would be an outstanding mutation, well worth the effort if properly done. Green hybrids especially in the first cross tend to show a more rust coloured shoulder patch than the pure Alexandrine.

◀ Hen Alexandrine.

▼ Cock Alexandrine.

Lutino hen 3/4 Alexandrine (flying 1 week). Note rust coloured shoulder patches on hybrid split Lutino cock. ▼

▲ Pair of Alexandrines.

▼ Cock Alexandrine.

Hen Alexandrine. ▶

▼ Alexandrine eggs.

Plum-headed Parrot

Psittacula cyanocephala cyanocephala

The Plum-heads just came into our collection of parrots in 1962 and ever since then have been a firm favourite.

Being a bird of only 33-35cm (13-14in) in length makes it an ideal size that isn't noisy, doesn't chew the normal bird netting and is an extremely attractive and colourful bird.

They have been well named, for if you were to describe the colouring of an adult male, the most striking feature is the head which is a very rich deep red to plum colour being separated from the green body colour by a distinct black band right around the neck. A chestnut shoulder patch is also present. Females are rather duller with the body green and the head grey, a yellowish band around the neck and lacking the chestnut shoulder patch.

Sexing young birds is extremely difficult. At what age these birds can be sexed depends on how early in the season the youngsters fledge. Young that fly early in November often show signs of colour when two years of age but don't fully colour until three years old. However, young that fledge late December early January don't show any signs of colour at all until their third year and then within weeks the cocks completely colour. If you know the age of the birds at least after 36 months they will be either cocks or hens.

Surgical sexing has been a great break through for aviculture and vets have become more and more experienced in this field, so breeders or buyers can get young sexed to save the long wait of wondering what one has.

Once a pair is obtained, we find them to be reasonably reliable breeders adapting to either a hot or cold climate. Originating from a hot country, we find Plum-heads have done exceptionally well in the colder climate of southern Victoria, not showing any signs of discomfort during the foggy, misty days that Melbourne often experiences.

Housing for these birds is fairly simple. As mentioned before being only a small bird they only require an aviary 3.00-3.60m (10-12ft) long, 90cm (3ft) wide and 2.10m (7ft) high, a third of the length should be shelter. Framing can be of wood as the birds do not chew a lot and either 12mm (1/2in) bird netting or weld mesh can be used on the open flight.

Plum-heads are best kept a single pair per flight but we have kept Bourkes with them with no problems, also in planted aviaries they really show themselves off flying amongst the greenery and not interfering with finches or doves.

The breeding of Plum-heads depends on compatibility of the pairs as we have found some hens are aggressive towards their partners, often taking control of the perch and bullying the male until a few weeks before egg laying. This often results in poor fertility. However, in compatible pairs hens don't do this at all, forming a pair bond with feeding and mating taking place many times before egg laying.

Nest logs or boxes can be used and we have found they prefer a small log with an inside diameter of around 15cm (6in) and about 38cm (15in) long with a small entrance hole 25-37mm (1.5-2in).

A suitable box size would be 18cm x 18cm x 45cm (7in x7in x 18in).

Plum-heads have a bad habit of removing sawdust from the nest log and it is surprising how quickly they are able to do this. Weekly inspections should be made to check that all sawdust is not removed, otherwise the hen will start laying on the bare bottom of the log. If a hen keeps carting the material out it is just a matter of replacing it with more moist hardwood sawdust, ramming this down tightly with your fist until she stops doing it. Once egg laying commences, they seem to be satisfied with the nesting chamber and rarely cart any more out.

The normal clutch usually consists of four to five eggs and incubation lasts about 23 days. All the young hatch together. The parents are very devoted to them with the hen leaving the nest to be fed by the cock regularly through the day then returning to feed the youngsters. Plum-heads had a reputation for letting young die after about three weeks of age but this has been overcome by keeping up a daily supply of boiled corn and seeding grasses along with their regular supply of dry seed consisting of grey sunflower, budgie mix and oats.

On leaving the nest, some seven weeks after hatching, young are a very plain green with pale yellow beaks not looking anywhere near as colourful as their parents. Going into a moult at around three months of age with the green head changing to a silvery grey colour. They retain this colour for at least two years and from then on cocks can show odd red feathers, either on the head or where the wing patch is.

Adult hens over eight to ten years sometimes show a few red feathers on the head, the numbers increasing as they grow older and inexperienced people may think they are young cocks just colouring up. However, with adults the beak of the hen is a much duller orange yellow than the cocks.

Plum-heads rarely have a successful second brood. We have tried fostering the first clutch of eggs under Indian Ringnecks and they will often lay again but infertile eggs often result so it appears to be a wasted exercise, just exhausting the hens.

◀ *Hen Plum-head.*

▼ *Young cock Plum-head breaking colour on face*

▲ *Cock Plum-head.*

Hen Plum-head. ▶

Hen Plum-head ▶ in her log. Note how she curls her tail over her back.

This Yellow or Pied ▶ cock Plum-head has got yellower over the years and is thus termed an acquired colour.

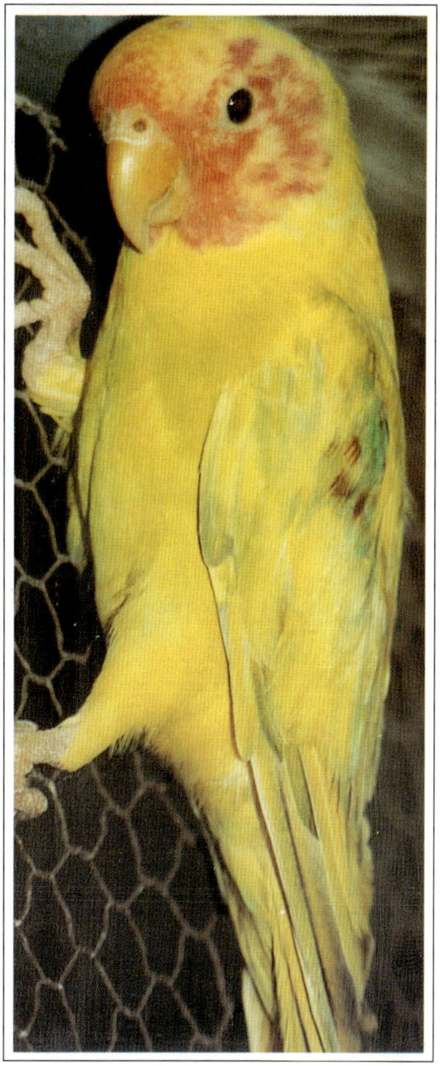

▼ "Uncoloured" young Plum-head

We find Plum-headed Parrots long lived with pairs in our collection breeding regularly for the past fifteen years. One of our cocks is at least twenty four years old and a hen at least twenty years old.Their numbers are steadily increasing in this country. Being so popular they are often exported to New Zealand where they also breed successfully.

Mutations of this Asiatic in Australia are rather rare. We have a male that has gradually over the years turned yellow still retaining the red head. This is what is termed an acquired colour. Efforts have been made to produce with him with little success.

Plum-heads are rather active birds in an aviary and if small twiggy branches are provided, cocks delight in sitting on the thinnest end bobbing up and down displaying to their hens. A magnificent sight cementing their future further into our collection

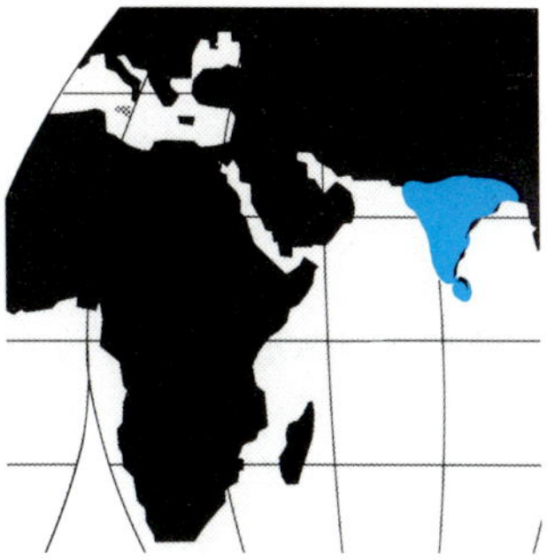

Indian Ringnecked Parrot

Psittacula krameri manillensis

Of the many types of parrots available to aviculture in Australia, Indian Ringnecks would be one of the most popular birds represented in parrot collections, certainly the most common of the foreign parrots. They have proved to be extremely hardy and reliable breeders.

Looking back over some 30 years of keeping this species, there are not many parrots that we could compare with this lovely Asiatic. Whether it be a dry, wet, hot or cold season it doesn't seem to affect their breeding performance. We have seen them housed under all types of conditions, from dark dingy aviaries to large spacious cages completely open with no protection from the weather but they always battle on. Certainly a lot of our native birds wouldn't last long if they had to live under similar conditions.

Being such free breeders and long livers, one would wonder that they haven't saturated their own market. However, having raised over 800 in the last 30 years, I find there is still a strong demand for them, especially the mutations which are often exported out of the country.

The 42cm (16.5 in.) normal Green Ringneck is a handsome bird especially when displaying to his female partner, stretching and bowing with eyes dilating. Noises like motor vehicles, lawn mowers etc passing the aviary usually set cocks off into a display and with uncoloured birds this is often a guide to their sex. Only cocks display.

Like most Asiatics, they are extremeley hard to sex under two to two and a half years of age unless you have them surgically sexed - otherwise it is a waiting game until black and pink feathers start to appear forming a ring around the neck of the cock birds. Hens remain completely green showing just an indentation where the neck ring should be. Adult plumage is attained at three years.

Indian Ringnecks look their worst when moulting and if you aren't familiar with them you would possibly wonder if something was dramatically wrong with their feathers. Within days, tail feathers drop and pin feathers appear everywhere and they look total wrecks. However, within a few weeks, new feathers replace the old and they are back to their immaculate condition. It is surprising how quickly cocks colour during this moulting period. One week no colour then completely coloured the next.

People that can't breed Ringnecks in our opinion, either haven't got a true pair or aren't trying. Even if nest logs are not provided during the breeding season, we have seen hens digging holes in the earth floor in a desperate bid to form a nesting chamber.

Single pairs are housed in open flights, one third being covered, varying in overall length from 4.80m to 6.60m (16ft to 22ft) but experience has proved 4.80m (16ft) is ample as the longer the flight the more speed youngsters get up when first

◀ *Hen Indian Ringneck.*

▼ *Cock Indian Ringneck.*

◀ *Albino hen Ringneck in nesting log with day old young.*

▶ Baby Ringnecks about four days old. Note red eye of Lutino.

▲ Ringneck eggs.

▲ Baby Ringnecks about ten days old (Note three Lutinos).

▲ Baby Ringnecks about sixteen days old.

▼ Baby Ringnecks about five weeks old.

leaving the nest. If a branch of leaves or something similar is not placed either end of the flight, damage to heads or wings can result. Youngsters soon learn the limits of the flight and thereafter there is no problem.

Hollow logs hung vertically are used for nesting. These measure 90cm (3ft) long with an inside diameter of 20-25cm (8-10in) and a 7.5cm (3in) entrance hole is made 15cm (6in) from the top. An inspection opening is made 23cm (9in) from the bottom. These inspection openings are certainly a time saver as regular checks are made on incubating progress and it is easy to take young out to be closed rung without removing the log from the wall. Nest boxes should be about 23cm (9in) square and about 60cm (2ft) high with a 7cm (3in) entrance hole.

Four eggs would be a normal clutch, however, five is not unusual with incubation taking 21 - 24 days and the young spending seven weeks in the nest. I find the Ringnecks excellent parents, rarely letting young die.

Many years ago, we had an experience with a normal pair of Ringnecks. A few days after hatching a swarm of bees tried to take over the nest log but the parents defended their territory and after a few cans of fly spray and many stings to ourselves, rearing continued with no ill effects.

On fledging, the young are often as big in the body as their parents, just being a bit shorter in the tail. But within a few months it is hard to distinguish them from the adult female.

Our Ringnecks are fed the basic mix of 2 parts grey sunflower, 1 part budgie mix, 1 part oats plus either boiled corn or corn on the cob when young are in the nest. This corn being continued until they lose interest in feeding on it, usually about two months after flying.

Green food is provided daily with milk thistle and/or oats in head and these are completely consumed stalk and all.

Ringnecks seem to eat anything from seeds, most types of fruit and vegetables and will be the first to try such commercial products as dry dog food or pellets. We believe they would find food and survive where most other birds would starve.

One big attraction with Ringnecks is the mutations that are available to aviculturists. Many people are against the establishment of mutations basically on the argument that the offspring is degrading from the original species. We can agree with such statements when many so called mutations often have problems such as: poor eyesight, poor feathering, slow breeding performance and sometimes one either needs a wild imagination to see the colour difference or they are just poor specimens, being of no value to aviculture. On the other hand, there are some excellent mutations that have proved themselves worthy and are well established in collections with Lutino and Blue Ringnecks rating high on this list.

In the late fifties, we had the privilege of viewing the vast parrot collection of Edward Hallstrom of Sydney. This ranged from many species of Macaws, African Greys, Lories, Eclectus etc. But of all these exotic parrots, the birds that appealed to us most were a large colony of Lutino Ringnecks. These birds are completely yellow, brighter on the face and wing margins with the neck ring of the males being pink and white instead of pink and black as in the normals. This was a sight we will always remember. Soon after that visit we were able to obtain some adult pairs of these magnificent birds which formed the nucleus of our breeding programme. Prior to that most Lutinos in private collections were very poor quality and much work was needed to improve this mutation.

Being sex-linked, it was a slow process to produce split cocks through normal green hens, but along with a strict culling programme over the years we are proud to say that today, Lutinos are as strong as the normals and are an excellent mutation, praised and admired by many overseas visitors.

Blue Ringnecks are off to a better start than Lutinos simply because they are recessive and the strength is retained as you are able to produce splits, either male or female from a normal.

The late Mr Joe Mattinson of Woolongong spent some twenty years developing a blue strain but unfortunately all his birds were disposed of after his death. It is amazing how quickly a collection can be broken up and if we are not careful, within a few years a species can be lost from aviculture forever. However, with careful management and provided accurate records are kept, breeders that have Blues should be able to increase the numbers of this beautiful mutation in the coming years. I think the Blue is as nice as the Lutino with the complete body colour being a soft powder blue brightest on the face with the male's neck ring black and white. I don't know if it is an illusion or just the contrast with the blue face but their red beaks seem to be brighter as if lacquered with nail polish. Indeed a very handsome bird.

The Grey Ringneck is also a very striking bird. Its three tone colours of silver, grey and black make it look very distinguished. The beak is red like the other Ringnecks. The Grey Ringneck was first bred in Australia by Mr Ian Adcock of New South Wales. The Grey is a dominant mutation and when paired with a normal Green Ringneck (the original dominant colour) produces the Grey-green Ringneck. This bird is a khaki green colour with black flight and tail feathers.

Albinos are well established in Australia having been bred through Blues and Lutinos. These birds are pure white which offsets their red beaks and eyes. Dilutes and Pieds are now also being developed.

Various colour matings will produce results that promise some exciting years ahead for Ringneck breeders.

Grey Ringneck in flight. ▶

Grey Indian Ringneck.

Blue and Grey Ringnecks.

Cock dilute Grey-green Ringneck.

Cock Grey-green Ringneck.

Grey Indian Ringneck.

Pied Ringneck.

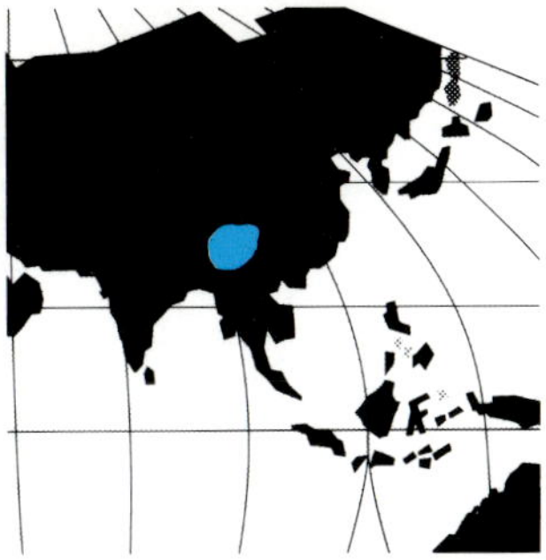

Derbyan Parrot

Psittacula derbyana

Having only seen photographs and paintings in bird books of Derbyans our minds were completely changed the first time we saw one in the flesh. We immediately went in search of obtaining some.

The feature that took our eye was the deep purple breast of these large birds which is a colour you don't see on many parrots. They also have a strong head with an alert eye and this seems to really set them off. Although they have the same body size as an Alexandrine, the tail doesn't grow as long which gives them a stocky appearance. The overall length would be approximately 50cm (20in).

Derbyans spend a lot of time chewing and perches have to be replaced regularly and an aviary of strong construction is needed with good type weld mesh.

Our Derbyans are housed in single pairs in open flight aviaries 5.40m (18ft) long, 2.10m (7ft) high and 1.20m (4ft) wide, a third of this being shelter. Some aviculturists haven't had a lot of success with this Asiatic but we have found them to be regular breeders, preferring a cold climate to a hot one.

A 90cm (3ft) log should be provided some 25-30cm (10-12in) inside diameter with a large opening of at least 13cm (5in) as these birds nest some six weeks later than other Asiatics and by the time young are ready to leave the nest we are starting to get hot weather with temperatures during the day up to 38C (100F). If there is not enough air circulation in the nest, young can become dehydrated.

Most Asiatics are born bald but not so with Derbyans. After 24 days of incubation young hatch with yellowish down, much like a homing pigeon, then this is replaced with a dense dark grey down like Eclectus get, giving them the appearance of a baby penguin.

If temperatures start to rise and you get concerned about youngsters, regular inspections should be made on days over 29C (85F). One can put the young in an open topped box on the floor of the aviary through the day, returning them to the log at night with no problems. Derbyans are most tolerant of this interference.

This Asiatic has an advantage in the fact that at eight months of age, you can sex them. Young leave the nest when about eight weeks old. They are as large as their parents but duller in colour with red beaks. At three months the beak colour changes to black and then at eight months cock birds get a red beak and the hens remain black.

One also gets an indication of sex by the presence of a pink line on the neck when the young are in the log, cocks lack this colour. However, some hens don't have as much pink as others so we only use this as a guide, but certainly the beak colour is more accurate.

We find the feeding of Derbyans unlike other parrots, a big

problem. We have had arguments with other breeders over sunflower seed. Some say it is a drug with birds becoming addicted to it and it also causes feather problems. We totally disagree with this for if you didn't feed sunflower to this species you would soon have some very dead Derbyans.

We have tried all types of parrot mixes and grey sunflower is virtually all we can get them to regularly eat. Even with green feed they are not too keen but they will take apples or either boiled corn or fresh corn cobs hung up in the aviary. After generations of breeding Derbyans, we have tried to entice young to take more greens but with little success.

Only on one occasion have we seen a poor feathered Derbyan. This bird had been hand raised and obviously hadn't received the right rearing mixture as it was undersized and a poor specimen all over.

Fortunately Derbyans are on the increase in this country and although still expensive good parrot collections are usually represented with this species.

Not usually breeding until three years old and with a small clutch size of two to three eggs nobody will mass produce them. We have pairs over twenty years old breeding regularly so they do have a good lifespan.

We know of hybrids with Alexandrines that have been raised but we don't know if the offspring are fertile, hopefully not. We are very fortunate to have Derbyans in Australia and it is up to aviculturists to keep the species true to type as we have been told they have decreased in their native land of Tibet and are also fairly rare in collections overseas.

▲ *Pair of Derbyans.*

▲ *Hen Derbyan at nest entrance.*

▼ *Pair of Derbyans - cock has red beak.*

◀ *Young Derbyan flying 2 days - note red beak at this stage.*

▼ *Baby Derbyan about four weeks old - note heavy down.*

▲ *Baby Derbyans about six weeks old.*

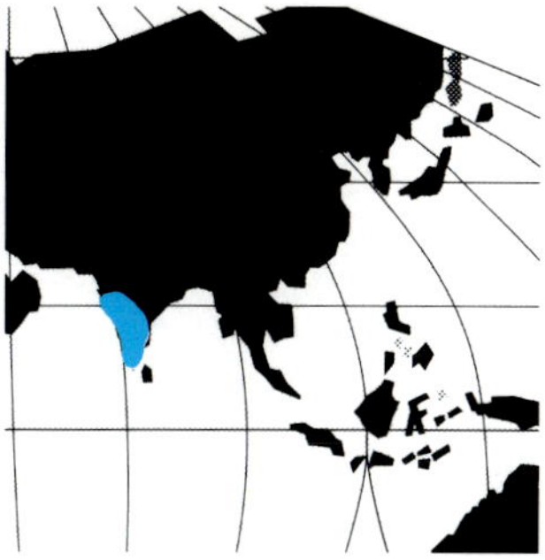

Malabar Parrot

Psittacula columboides

This is an Asiatic that we haven't kept at Harkaway. There is a small population in Australia that are kept by a few dedicated fanciers. Their main body colour is a bluish grey with the back feathers laced with black. Like all Asiatics, a distinct black ring encircles the neck and directly behind this, a turquoise green colour which increases with age only on males. They are about 38cm (15in) in length.

Young are a duller version of their parents leaving the nest with black beaks but at eight to ten months red replaces the black as the young cocks start to colour, the hen's beak remains black. Adult plumage is acquired at two years and from this age on displaying and nesting can be expected.

Speaking to breeders, a point of interest with hand raised young is that they remain tame when put out in an aviary, other Asiatics hand raised soon lose their tameness unless kept in a small cage. Even parent-reared young are much quieter than other Asiatics and soon learn to fly to the front wire when approached looking for tid-bits.

Being much the same size as a Slatey-head, their requirements are the same for feeding, housing and nesting.

The late Mr Joe Mattinson persevered with this Asiatic for many years and his were kept in single pairs. Their incubation period is about 24 days and the young remain in the nest for seven weeks. His best breeding pairs were badly feather-plucked, much the same way as some Northern Rosellas become in the breeding season so Indian Ringnecks were used as foster parents which did an excellent job rearing the young.

The feather problem doesn't seem to be as bad with birds that are housed in planted aviaries, the larger the better.

Fortunately, a very experienced aviculturist procured most of Mr Mattinson's stock.

Let's hope from now on, with the help of vets and a bit of understanding we will see Malabars in many more parrot collections - as these birds are rather rare world-wide.

Pair of Malabars - cock has red beak. ▶

◀ Hen Malabar.

▼ Baby Malabars about ten days old.

◀ Cock Malabar.

▼ Cock Malabar.

▲ Malabars are probably the quietest of the Asiatics.

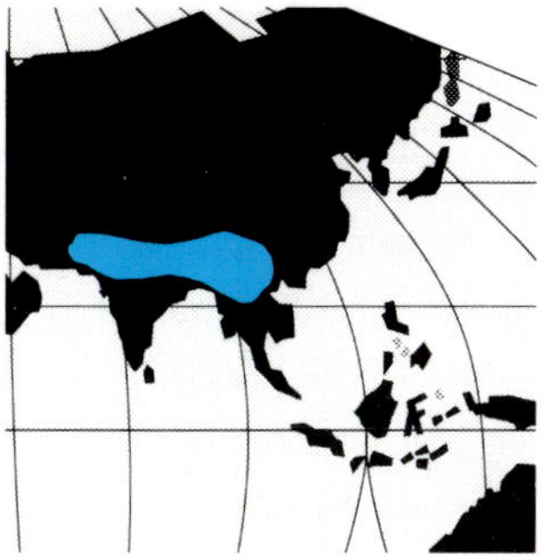

Slatey-headed Parrot

Psittacula himalayana himalayana

Slatey-heads have only been added to our collection in the last few years. The reason they were absent was we could never compare them to being as nice as a Plum-head, however, repeated enquiries for them has led to us keeping a few pairs.

When in peak condition they are a nice bird about 40cm (16in) in length. The most striking feature is their long tail feathers which gives them a stream-lined, slender appearance but if they are housed in small aviaries the tails often get broken and this detracts from their appearance.

In Melbourne, Slatey-heads look their best around July with both cocks and hens showing a deep dark slate grey on the head with bright red bills.

We find that as the hens are about to commence nesting around August, the head turns a duller grey and the beaks have a chalky appearance. Both cock and hen are similar in appearance except hens don't carry the maroon wing patch.

Housing for this species is similar to that of the Plum-head but being slightly larger, a bigger log is required about 20cm (8in) in diameter and 45cm (18in) to 60cm (24in) long. Boxes 20cm (8in) square and 45cm (18in) to 60cm (24in) high are ideal. Hens settle down quicker than Plum-heads, not removing sawdust from the log and often adding down feathers once egg laying commences. They are also very tolerant when their nest is inspected.

Slatey-heads have a good reputation for rearing their own young right through without any hassles, not usually needing to be taken and hand-reared or fostered under other species.

Clutches of four to five eggs are not uncommon. Incubation lasts about 24 days and young stay in the nest for around seven weeks. On leaving they are rather uninterestingly green but like Plum-heads, go into a moult and heads change to grey at around six months, often at that age odd black feathers appear on the neck ring and at fifteen to eighteen months one is able to sex cocks from hens. Adult plumage isn't acquired until two to two and half years of age.

Feeding these birds is the same as with all other Asiatics using our basic seed mix and plenty of milk thistle and corn on the cob when they have young. Slatey-heads will readily eat fruit. On weaning, young are put together in a holding aviary where they show no signs of aggression towards one another.

▼ *Pair of Slatey-heads - cock on left.*

▲ *Pair of Slatey-heads with their young.*

▼ *Hen Slatey-head in nesting log with young.*

▲ *Hen Slatey-head at log entrance.*

▼ *Pair of Slatey-heads.*

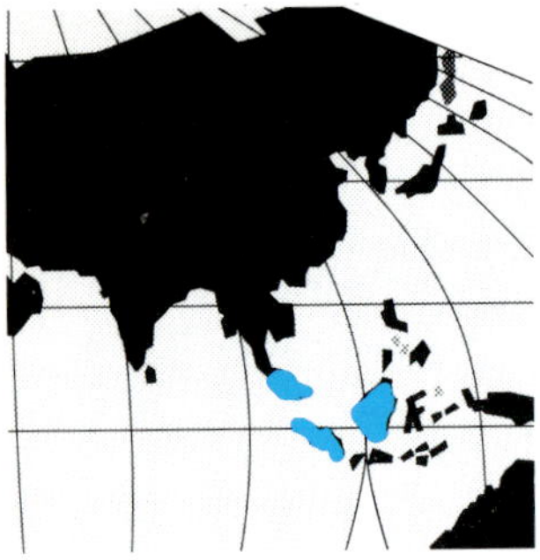

Malayan Long-tailed Parrot

Psittacula longicauda

Malayan Long-tails came into our collection in the early 1960's when four uncoloured birds were purchased from a Melbourne dealer. The major difference we found in the management of these birds compared to other Asiatics was that they required a much larger daily supply of fruit. Other than that we found their requirements to be similar to the Slatey-headed Parrot.

In adult plumage the males live up to their name as they have two distinct very long blue tail feathers which look out of proportion to the bird's body size. The hen's tail is a much more sensible length. The cock is predominantly green with a thick black moustache and red cheeks. The beak is red. The female has less red cheeks, a green moustache and the beak is black. The cock is about 40cm (16in) in length. Like the Derbyan they have a small clutch size of two to three eggs. Incubation lasts about 24 days.

Looking back now, we regret having exchanged our Long-tails for other birds. There are not too many left in Australia at the moment and we can only hope that through intensive management and careful use of the gene pool we have available these birds can be saved for future Australian aviculturists to enjoy.

Pair of Malayan Long-tails - cock right.

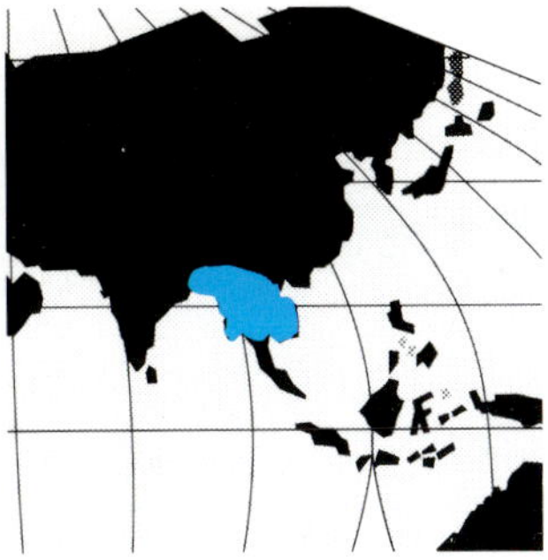

Blossom-headed Parrot

Psittacula cyanocephala rosa

The Blossom-headed Parrot is extremely similar to the Plum-headed Parrot and their requirements are identical.

There are still a number of Blossom-heads in this country which are held by a few experienced aviculturists so hopefully they will increase in numbers.

Blossom-heads are slightly smaller than a Plum-head - about 30cm (12in) in length - with the main difference being that cocks are more of a pink colour on the head, not getting the plum colour of Plum-heads. Females are similar to hen Plum-heads but have a maroon patch on the wing which is lacking in the hen Plum-heads, also both birds tails are tipped with yellow, this is white in Plum-heads.

Possibly the reason Blossom-heads are not as common in aviaries today is the price difference between them and the Plum-heads. A lot of people go for value and colour in a species. A comparison could be made with Cloncurry Parrots and Mallee Ringnecks for instance. Many people cannot understand how similar looking birds can vary so much in price.

Blossom-heads will readily cross with Plum-heads and it would take a trained eye to detect these hybrids. Hopefully breeders will retain the species true.

▲ *Cock blossom-head.*

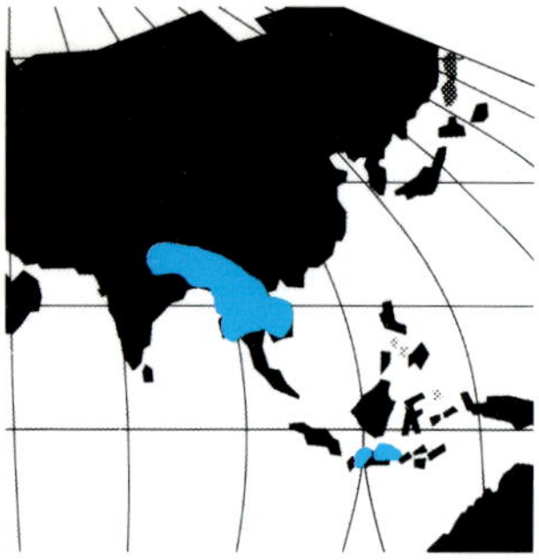

Moustached Parrot

Psittacula alexandri fasciata

Purchasing all stock from the late Mr Joe Mattinson's collection in April 1988, our experience with this species to date is rather limited. However, in this short time, we find they are very similar to a Derbyan except they are only half the size and thankfully eat all we put before them.

When being caught they roll over in the hand net on their backs squeaking continually exactly as Derbyans do.

Another point of interest we have noticed is that after the breeding season hens hide in the nest log until about July when they give this unusual behaviour away, spending more time out on the perch. We have no explanation to date, as to why they do this. Rarely do cocks go into the logs.

Sexing of Moustached is the same as Derbyans - the young cock's beak colour changes to a bright red at eight months of age. The hen's remains black.

Moustached have rather a plump body with a reasonably short tail compared with other Asiatics, total length would be about 33cm (13in). Cocks and hens are very similar in colour, the main difference being beak colour - red in cocks and black in hens.

Large aviaries are not required and ours are housed a single pair per flight 3.00-3.60m (10-12ft) long, 90cm (3ft) wide and 2.10m (7ft) high, a third of this being shelter seem to be adequate. Though not chewing as much as Derbyans, we still use a heavy gauge netting.

Nest logs hung vertically measuring 60cm (2ft) long with an inside diameter of 20cm (8in) and an entrance hole of 5cm (2in) we find ideal.

Hens spend a lot of time in the log before egg laying commences. However, once settled down, they lay a clutch of three to four eggs which are rather large for the size of the bird. Hens sit tight, only leaving the nest early in the morning or late in the afternoon to feed herself or be fed by the cock.

On nest inspections, they are very hard to persuade to move off their eggs. We would estimate incubation at around 21-24 days. The short experiences we have had has resulted in the parents being excellent parents with the young growing rapidly over the seven weeks they are in the nest. On leaving, young are sensible, not flying erratically as Asiatic youngsters often do.

Beak colour is pink but within weeks changes to black until at eight months of age the cocks achieve red beaks. Full adult plumage is attained after fifteen months.

Feeding habits are rather different from Derbyans as although they eat large amounts of grey sunflower, they will also take the smaller seeds in the budgie mix. Oats are eagerly consumed when young are in the nest. Green feed in the form of milk thistle and oats in head hung up in the flight are a great favourite.

There is also a sub-species of Moustached *P. a. alexandri* in Australia. The hens are identical to cocks except for a faint pink line on the neck directly behind the black moustache. This race is rather aggressive towards one another. The more common race we have *P. a. fasciata* are very compatible towards each other and we are able to leave the young with their parents until they colour.

▲ *Cock Moustached.*

▼ *Hen Moustached.*

◀ *Cock Moustached.*

▼ *Cock Moustached.*

A striking comparison between cock Derbyan (left) and cock Moustached.

Hen Moustached.

Cock Moustached with young (flying one week) - note young's bill is red at this stage.

Baby Moustached about three days old.

Genetics and Colour Breeding

From past experience we have found that many people do not want to know about the actual mechanics of genetic inheritance. The expected progeny results of a certain pairing is all they are really interested in. So for all Indian Ringneck mutation breeders, here is a list of some matings and their expectations.

The genetic inheritance of the various mutations are as follows:-

Blue	-	Recessive
Lutino	-	Sex-linked
Albino	-	Sex-linked
Grey	-	Dominant

Grey Ringnecks are blue series birds. All single factor Grey or Grey-green birds can produce Blues when paired to Blues, Green/Blues or Albinos.

The dominant Grey and Grey-green genes can occur in Single Factor (SF) and Double Factor (DF) forms. Their mode of inheritance is as follows:-

SF x Normal	=	50% SF	50% Normal	
SF x SF	=	50% SF	25% DF	25% Normal
SF x DF	=	50% SF	50% DF	
DF x Normal	=	100% SF		
DF x DF	=	100% DF		

Parents Cock x Hen	%	Progeny Male	%	Female
Green x Green/Blue	25% 25%	Green Green/Blue	25% 25%	Green Green/Blue
Green x Lutino	50%	Green/Lutino	50%	Green
Green x Lutino/Blue	25% 25%	Green/Lutino Green/Blue/Lutino	25% 25%	Green Green/Blue
Green x Blue	50%	Green/Blue	50%	Green/Blue
Green x Albino	50%	Green/Blue/Lutino	50%	Green/Blue
Green/Lutino x Green	25% 25%	Green Green/Lutino	25% 25%	Green Lutino
Green/Lutino x Green/Blue	12.5% 12.5% 12.5% 12.5%	Green Green/Lutino Green/Blue Green/Blue/Lutino	12.5% 12.5% 12.5% 12.5%	Green Green/Blue Lutino Lutino/Blue
Green/Lutino x Lutino	25% 25%	Green/Lutino Lutino	25% 25%	Green Lutino
Green/Lutino x Lutino/Blue	12.5% 12.5% 12.5% 12.5%	Green/Lutino Green/Blue/Lutino Lutino Lutino/Blue	12.5% 12.5% 12.5% 12.5%	Green Green/Blue Lutino Lutino/Blue
Green/Lutino x Blue	25% 25%	Green/Blue Green/Blue/Lutino	25% 25%	Green/Blue Lutino/Blue
Green/Lutino x Albino	25% 25%	Green/Blue/Lutino Lutino/Blue	25% 25%	Green/Blue Lutino/Blue
Green/Blue x Green	25% 25%	Green Green/Blue	25% 25%	Green Green/Blue
Green/Blue x Green/Blue	12.5% 25% 12.5%	Green Green/Blue Blue	12.5% 25% 12.5%	Green Green/Blue Blue
Green/Blue x Lutino	25% 25%	Green/Lutino Green/Blue/Lutino	25% 25%	Green Green/Blue

Green/Blue x Lutino/Blue	12.5%	Green/Lutino	12.5%	Green
	25%	Green/Blue/Lutino	25%	Green/Blue
	12.5%	Blue/Lutino	12.5%	Blue
Green/Blue x Blue	25%	Green/Blue	25%	Green/Blue
	25%	Blue	25%	Blue
Green/Blue x Albino	25%	Green/Blue/Lutino	25%	Green/Blue
	25%	Blue/Albino	25%	Blue
Green/Blue/Lutino x Green	12.5%	Green	12.5%	Green
	12.5%	Green/Lutino	12.5%	Green/Blue
	12.5%	Green/Blue	12.5%	Lutino
	12.5%	Green/Blue/Lutino	12.5%	Lutino/Blue
Green/Blue/Lutino x Green/Blue	6.25%	Green	6.25%	Green
	12.5%	Green/Blue	12.5%	Green/Blue
	12.5%	Green/Blue/Lutino	12.5%	Lutino/Blue
	6.25%	Blue	6.25%	Blue
	6.25%	Blue/Albino	6.25%	Albino
	6.25%	Green/Lutino	6.25%	Lutino
Green/Blue/Lutino x Lutino	12.5%	Green/Lutino	12.5%	Green
	12.5%	Green/Blue/Lutino	12.5%	Green/Blue
	12.5%	Lutino	12.5%	Lutino
	12.5%	Lutino/Blue	12.5%	Lutino/Blue
Green/Blue/Lutino x Lutino/Blue	6.25%	Green/Lutino	6.25%	Green
	12.5%	Green/Blue/Lutino	12.5%	Green/Blue
	6.25%	Lutino	6.25%	Lutino
	12.5%	Lutino/Blue	12.5%	Lutino/Blue
	6.25%	Blue/Albino	6.25%	Blue
	6.25%	Albino	6.25%	Albino
Green/Blue/Lutino x Blue	12.5%	Green/Blue	12.5%	Green/Blue
	12.5%	Green/Blue/Lutino	12.5%	Blue
	12.5%	Blue	12.5%	Lutino/Blue
	12.5%	Blue/Albino	12.5%	Albino
Green/Blue/Lutino x Albino	12.5%	Green/Blue/Lutino	12.5%	Green/Blue
	12.5%	Lutino/Blue	12.5%	Lutino/Blue
	12.5%	Blue/Albino	12.5%	Blue
	12.5%	Albino	12.5%	Albino
Lutino x Green	50%	Green/Lutino	50%	Lutino
Lutino x Green/Blue	25%	Green/Lutino	25%	Lutino
	25%	Green/Blue/Lutino	25%	Lutino/Blue

Lutino x Lutino	50%	Lutino	50%	Lutino
Lutino x Lutino/Blue	25% 25%	Lutino Lutino/Blue	25% 25%	Lutino Lutino/Blue
Lutino x Blue	50%	Green/Blue/Lutino	50%	Lutino/Blue
Lutino x Albino	50%	Lutino/Blue	50%	Lutino/Blue
Lutino/Blue x Green	25% 25%	Green/Lutino Green/Blue/Lutino	25% 25%	Lutino Lutino/Blue
Lutino/Blue x Green/Blue	12.5% 25% 12.5%	Green/ Lutino Green/Blue/Lutino Blue/Albino	12.5% 25% 12.5%	Lutino Lutino/Blue Albino
Lutino/Blue x Lutino	25% 25%	Lutino Lutino/Blue	25% 25%	Lutino Lutino/Blue
Lutino/Blue x Lutino/Blue	12.5% 25% 12.5%	Lutino Lutino/Blue Albino	12.5% 25% 12.5%	Lutino Lutino/Blue Albino
Lutino/Blue x Blue	25% 25%	Green/Blue/Lutino Blue/Albino	25% 25%	Lutino/Blue Albino
Lutino/Blue x Albino	25% 25%	Lutino/Blue Albino	25% 25%	Lutino/Blue Albino
Blue x Green	50%	Green/Blue	50%	Green/Blue
Blue x Green/Blue	50% 25%	Green/Blue Blue	50% 25%	Green/Blue Blue
Blue/Lutino	50%	Green/Blue/Lutino	50%	Green/Blue
Blue x Lutino/Blue	25% 25%	Green/Blue/Lutino Blue/Albino	25% 25%	Green/Blue Blue
Blue x Blue	50%	Blue	50%	Blue
Blue x Albino	50%	Blue/Albino	50%	Blue
Blue/Albino x Green	25% 25%	Green/Blue Green/Blue/Lutino	25% 25%	Green/Blue Lutino/Blue

Blue/Albino x Green/Blue	12.5%	Green/Blue/Lutino	12.5%	Green/Blue
	12.5%	Green/Blue	12.5%	Lutino/Blue
	12.5%	Blue	12.5%	Blue
	12.5%	Blue/Albino	12.5%	Albino
Blue/Albino x Lutino	25%	Green/Lutino/Blue	25%	Green/Blue
	25%	Lutino/Blue	25%	Lutino/Blue
Blue/Albino x Lutino/Blue	12.5%	Green/Blue/Lutino	12.5%	Green/Blue
	12.5%	Lutino/Blue	12.5%	Lutino/Blue
	12.5%	Blue/Albino	12.5%	Blue
	12.5%	Albino	12.5%	Albino
Blue/Albino x Blue	25%	Blue/Albino	25%	Blue
	25%	Blue	25%	Albino
Blue/Albino x Albino	25%	Blue/Albino	25%	Blue
	25%	Albino	25%	Albino
Albino x Green	50%	Green/Blue/Lutino	50%	Lutino/Blue
Albino x Green/Blue	25%	Green/Blue/Lutino	25%	Lutino/Blue
	25%	Blue/Albino	25%	Albino
Albino x Lutino	50%	Lutino/Blue	50%	Lutino/Blue
Albino x Lutino/Blue	25%	Lutino/Blue	25%	Lutino/Blue
	25%	Albino	25%	Albino
Albino x Blue	50%	Blue/Albino	50%	Albino
Albino x Albino	50%	Albino	50%	Albino

In the following section on Grey Ringnecks:-

SF = Single Factor
DF = Double Factor

SF Grey x SF Grey	25%	SF Grey	25%	SF Grey
	12.5%	DF Grey	12.5%	DF Grey
	12.5%	Blue	12.5%	Blue
DF Grey x SF Grey	25%	SF Grey	25%	SF Grey
	25%	DF Grey	25%	DF Grey
DF Grey x DF Grey	50%	DF Grey	50%	DF Grey

SF Grey x Green	25% 25%	SF Grey-green Green/Blue	25% 25%	SF Grey- green Green/Blue
DF Grey x Green	50%	SF Grey-green	50%	SF Grey-green
SF Grey x SF Grey-green	12.5% 6.25% 12.5% 6.25% 6.25% 6.25%	SF Grey DF Grey SF Grey-green DF Grey-green Blue Green/Blue	12.5% 6.25% 12.5% 6.25% 6.25% 6.25%	SF Grey DF Grey SF Grey-green DF Grey-green Blue Green/Blue
DF Grey x SF Grey-green	12.5% 12.5% 12.5% 12.5%	SF Grey DF Grey SF Grey-green DF Grey-green	12.5% 12.5% 12.5% 12.5%	SF Grey DF Grey SF Grey-green DF Grey-green
DF Grey x DF Grey-green	25% 25%	DF Grey DF Grey-green	25% 25%	DF Grey DF Grey-green
SF Grey-green x SF Grey-Green	12.5% 6.25% 12.5% 6.25% 6.25% 6.25%	SF Grey DF Grey SF Grey-green DF Grey-green Green/Blue Blue	12.5% 6.25% 12.5% 6.25% 6.25% 6.25%	SF Grey DF Grey SF Grey-green DF Grey-green Green/Blue Blue
Df Grey-green x SF Grey-green	12.5% 12.5% 12.5% 12.5%	DF Grey SFGrey DF Grey-green SF Grey-green	12.5% 12.5% 12.5% 12.5%	DF Grey SF Grey DF Grey-green SF Grey-green
DF Grey-green x DF Grey-green	25% 25%	DF Grey DF Grey-green	25% 25%	DF Grey DF Grey-green
SF Grey-green x Green	12.5% 12.5% 25%	SF Grey-green Green Green/Blue	12.5% 12.5% 25%	SF Grey-green Green Green/Blue
DF Grey-green x Green	50%	SF Grey-green	50%	SF Grey/green
SF Grey x Blue	25% 25%	SF Grey Blue	25% 25%	SF Grey Blue
DF Grey x Blue	50%	SF Grey	50%	SF Grey

Pairing	%	Offspring	%	Offspring
SF Grey-green x Blue	12.5%	SF Grey	12.5%	SF Grey
	12.5%	SF Grey-green	12.5%	SF Grey-green
	12.5%	Green/Blue	12.5%	Green/Blue
	12.5%	Blue	12.5%	Blue
DF Grey-green x Blue	25%	SF Grey	25%	SF Grey
	25%	SF Grey-green	25%	SF Grey-green
SF Grey x Green/Blue	12.5%	SF Grey	12.5%	SF Grey
	12.5%	SF Grey-green	12.5%	SF Grey-green
	12.5%	Green/Blue	12.5%	Green/Blue
	12.5%	Blue	12.5%	Blue
DF Grey x Green/Blue	25%	SF Grey	25%	SF Grey
	25%	SF Grey-green	25%	SF Grey-green
SF Grey x Lutino	25%	SF Grey-green/ Lutino	25%	SF Grey-green
	25%	Green/Blue/Lutino	25%	Green/Blue
Lutino x SF Grey	25%	SF Grey-green/ Lutino	25%	Lutino/Blue *(Masking Green)*
	25%	Green/Blue/Lutino	25%	Lutino/Blue *(Masks SF Grey-green)*
Lutino x DF Grey	50%	SF Grey-green/ Lutino	50%	Lutino/Blue *(Masks SF Grey-green)*
SF Grey-green x Lutino *(Masking SF Grey-green)*	6.25%	Green/Lutino	6.25%	Green
	6.25%	Grey-green/Lutino	6.25%	Grey-green
	6.25%	Blue/Albino	6.25%	Blue
	6.25%	Grey/Albino	6.25%	Grey
	12.5%	Green/Blue/Lutino	12.5%	Green/Blue
	12.5%	Grey-green/Lutino	12.5%	Grey-green

Note:- Of the Greys and Grey-greens, 66.6% will be SF.33.3% will be DF. Test mate to find out which are which.

Pairing	%	Offspring	%	Offspring
Green/Lutino x Lutino *(Masking SF Grey-green)*	6.25%	Green/Lutino	6.25%	Green
	12.5%	SF Grey-green/ Lutino	12.5%	SF Grey-green
	6.25%	Green/Blue/Lutino	6.25%	Green/Blue
	6.25%	Lutino	6.25%	Lutino
	12.5%	Lutino *(Masks SF Grey-green)*	12.5%	Lutino *(Masks SF Grey-green)*
	6.25%	Lutino/Blue	6.25%	Lutino/Blue

Mating	Expectation		Expectation	
Lutino *(Masking SF Grey-green)* x SF Grey-green	6.25%	Green/Lutino	6.25%	Lutino
	18.75%	Grey-green/Lutino	18.75%	Lutino *(Masking Grey-green)*
	12.5%	Green/Blue/Lutino	12.5%	Lutino/Blue
	6.25%	Blue/Albino	6.25%	Albino *(Masking Blue)*
	6.25%	Grey/Albino	6.25%	Albino *(Masking Grey)*
	Note: 66.6%. SF. 33.3% DF. Test mate.			
SF Grey-green/ Lutino x Lutino (Masking SF Grey-green)	3.125%	Green/Lutino	3.125%	Green
	9.375%	Grey-green/Lutino	9.375%	Grey-green
	6.25%	Green/Blue/Lutino	6.25%	Green/Blue
	3.125%	Lutino	3.125%	Lutino
	9.375%	Lutino *(Masking Grey-green)*	9.375%	Lutino *(Masking Grey-green)*
	6.25%	Lutino/Blue	6.25%	Lutino/Blue
	3.125%	Blue/Albino	3.125%	Blue
	3.125%	Grey/Albino	3.125%	Grey
	3.125%	Albino *(Masking Blue)*	3.125%	Albino *(Masking Blue)*
	3.125%	Albino *(Masking Grey)*	3.125%	Albino *(Masking Grey)*
	Note: 66.6% SF. 33.3% DF. Test mate.			
Green/Blue Lutino x SF Grey-green	3.125%	Green	3.125%	Green
	9.375%	SF Grey-green	9.375%	SF Grey-green
	6.25%	Green/Blue	6.25%	Green/Blue
	6.25%	Green/Blue/Lutino	6.25%	Lutino/Blue
	9.375%	SF Grey-green/ Lutino	9.375%	Lutino *(Masks SF Grey-green)*
	3.125%	Blue	3.125%	Blue
	3.125%	SF Grey	3.125%	SF Grey
	3.125%	Blue/Albino	3.125%	Albino *(Masking Blue)*
	3.125%	SF Grey/Albino	3.125%	Albino *(Masking SF Grey)*
	3.125%	Green/Lutino	3.125%	Lutino
Blue x Lutino *(Masks SF Grey-green)*	25%	Green/Blue/Lutino	25%	Green/Blue
	25%	SF Grey-green/ Lutino	25%	SF Grey-green

Pairing	%	Cocks	%	Hens
Lutino *(Masking SF Grey-green)* x Blue	12.5%	Green/Blue/Lutino	12.5%	Lutino/Blue
	12.5%	SF Grey-green/ Lutino	12.5%	Lutino *(Masks SF Grey-green)*
	12.5%	Blue/Albino	12.5%	Albino *(Masking Blue)*
	12.5%	SF Grey/Albino	12.5%	Albino *(Masking SF Grey)*
SF Grey-green/ Lutino x Blue	6.25%	Green/Blue	6.25%	Green/Blue
	6.25%	SF Grey-green	6.25%	SF Grey-green
	6.25%	Green/Blue/Lutino	6.25%	Lutino/Blue
	6.25%	SF Grey-green/ Lutino	6.25%	Lutino *(Masks SF Grey-green)*
	6.25%	Blue	6.25%	Blue
	6.25%	SF Grey	6.25%	SF Grey
	6.25%	Blue/Albino	6.25%	Albino *(Masking Blue)*
	6.25%	SF Grey/Albino	6.25%	Albino *(Masking SF Grey)*
DF Grey-green/ Lutino x Blue	12.5%	SF Grey-green	12.5%	SF Grey-green
	12.5%	SF Grey-green/ Lutino	12.5%	Lutino *(Masks SF Grey-green)*
	12.5%	SF Grey	12.5%	SF Grey
	12.5%	SF Grey/Albino	12.5%	Albino *(Masking SF Grey)*
SF Grey/Albino x SF Grey-green	6.25%	Green/Blue	6.25%	Green/Blue
	6.25%	Grey-green	6.25%	Grey-Green
	6.25%	Green/Blue/Lutino	6.25%	Lutino/Blue
	6.25%	Grey-Green/Lutino	6.25%	Lutino *(Masking Grey-green)*
	6.25%	Grey	6.25%	Grey
	6.25%	Blue	6.25%	Blue
	6.25%	Blue/Albino	6.25%	Albino *(Masking Blue)*
	6.25%	Grey/Albino	6.25%	Albino *(Masking Grey)*
	Note: - 66.6% SF. 33.3% DF. Test mate.			
Lutino/Blue x SF Grey	12.5%	SF Grey	12.5%	Albino *(Masking SF Grey)*
	12.5%	Blue/Albino	12.5%	Albino *(Masking Blue)*
	12.5%	SF Grey-green/ Lutino	12.5%	Lutino/Blue *(Masking Green)*
	12.5%	Green/Blue/Lutino	12.5%	Lutino/Blue *(Masks SF Grey-green)*

Lutino/Blue x DF Grey	25% SF Grey/Albino 25% SF Grey-green/ Lutino	25% Albino *(Masking SF Grey)* 25% Lutino/Blue *(Masks SF Grey-green)*
Albino x SF Grey	25% SF Grey/Albino 25% Blue/Albino	25% Albino *(Masking SF Grey)* 25% Albino *(Masking Blue)*
Albino x DF Grey	50% SF Grey/Albino	50% Albino *(Masking SF Grey)*
Albino *(Masking SF Grey)* x SF Grey	25% SF Grey/Albino 12.5% DF Grey/Albino 12.5% Blue/Albino	25% Albino *(Masking SF Grey)* 12.5% Albino *(Masking DF Grey)* 12.5% Albino *(Masking Blue)*
Albino *(Masking DF Grey)* x Blue	50% SF Grey/Albino	50% Albino *(Masking SF Grey)*
Albino *(Masking Blue)* x Albino *(Masking SF Grey)*	25% Albino *(Masking Blue)* 25% Albino *(Masking SF Grey)*	25% Albino *(Masking Blue)* 25% Albino *(Masking SF Grey)*
Albino *(Masking SF Grey)* x Albino *(Masking SF Grey)*	12.5% Albino *(Masking SF Grey)* 6.25% Albino *(Masking DF Grey)* 6.25% Albino *(Masking Blue)*	12.5% Albino *(Masking SF Grey)* 6.25% Albino *(Masking DF Grey)* 6.25% Albino *(Masking Blue)*

We will stop here. Any further Grey and ino combinations can be worked out from looking at previous matings but it should be obvious that in the future Lutinos and Albinos will have to be test-mated in many cases to see whether they are masking single factor or double factor Grey, Grey-green or Blue.

Pied Ringnecks are being developed in Australia but their mode of inheritance here has not been established at present.

Diseases of Asiatic Parrots

An Introduction

The Asiatics are generally more resistant to diseases compared to the Australian species of parrots, but this does not mean these birds are totally immune to disease. Poor management practices or exposure to highly virulent diseases will cause serious illness and even death, in all species, including the Asiatics.

1. Health check of the newly acquired bird

The most important aspect in maintaining the health of Asiatics is to acquire a bird already in good health. The emphasis here is the health of the new bird (rather than the risk of disease entering your aviary) because most Asiatic Parrots represent quite a sizeable investment for any aviculturist. This investment is best protected by a complete physical examination before purchase and a health check after purchase (with the full knowledge of the vendor). The health check must be done by an avian veterinarian with expertise in parrots. A diagnosis can be made at this time of any diseases which may affect the long term livelihood and breeding capabilities of the new purchase. Other less serious diseases may also be detected and appropriate therapy can be dispensed at this time to prevent problems. A bird diagnosed as having a serious disease which cannot be cured with appropriate medication or which will affect the breeding capabilities of the bird should not be purchased.

Nowadays, with the modern day diagnostic procedures it is possible to buy an expensive bird and guarantee its health at the time of purchase.

The health check benefits all aviculture by

(a) Providing a means of guarantee of an expensive bird.

(b) Removes the responsibility from the vendor to the buyer after a health check is completed satisfactorily.

(c) Discourages the unscrupulous sale of unfit birds to the unsuspecting aviculturist.

A complete physical examination of the newly acquired bird is necessary if a health check by an avian veterinarian is not taken. The most important part of the examination is to pick the bird up. A careful examination starts at the head, check the eyes, the nostrils, the feathers around the nostrils, the head, body, wings and especially the tail for signs of abnormal feathers, the vent for signs of pasting and the keel bone for "going light". The birds must be checked carefully for watery eyes and discoloured or malformed nostrils. These are signs of conjunctivitis and sinusitis. Left untreated for any period of time, these diseases often become impossible to cure. Both are symptoms of Chlamydiosis or a bacterial infection. Chlamydiosis is a particularly dangerous disease to birds as well as humans and you must be able to recognise its symptoms. Birds which have a prominent keel bone are said to be "going light". This has many causes, most of which are dangerous to your resident flock. Such a bird should not be purchased. It is very important to check the feathers. Many feather abnormalities seen in this group of birds indicate virus diseases such as Papovavirus or Feather and Beak Virus.

2. Quarantine of the new bird

The physical check of the new bird is the first step in preventing disease from entering your aviaries. The quarantine procedure is necessary to detect any diseased birds which passed the new bird physical check. Some birds appear healthy but in fact many carry a disease. These birds are called "carriers" and the quarantine procedure allows us to detect such birds. Many of these diseases are stress related and within the six week quarantine period most of these diseases will become apparent. The quarantine procedure is still recommended even with a veterinarian health check.

Quarantine gives the birds time to acclimatise to their new surroundings so that by the time they enter the aviary they are not in a stress situation. This lessens the chance of introducing disease to your aviary. It is important to place the birds into a quiet, secure quarantine aviary during the day and not at dusk so as to minimise the chance of wall crashing. During the quarantine time the Asiatic Parrots should be treated routinely for worms and mites using Ivermectin. Ivermectin is a restricted drug and is available only by prescription. The advantage in using Ivermectin is that it will kill both the worms and the mites in this group of parrots. It is important that the Asiatic Parrots are separated into compatible groups whilst in quarantine. It is also necessary to have a quarantine aviary with double wire to prevent aggression between aviary cages. Fighting in the quarantine aviary will stress the birds and make them prone to illness. The quarantine aviary should be away from the main aviary and should provide the new birds with a quiet and secure environment during their acclimatisation. It is best to have birds of the same or compatible species and from the same breeder in the quarantine aviary at the one time, so as to minimise the chance of spreading disease. It is important to carefully observe the new birds for signs of illness, especially signs of loose droppings, watery eyes and sick bird look (SBL). Any bird showing signs of illness must be immediately isolated. All healthy quarantined birds are allowed to enter the main aviary after six weeks. It is best to do this on a sunny weekend morning so that the birds can adjust easily to the new aviary. The quarantine aviary is then disinfected and cleaned in preparation for the next new purchase.

3. Stress

Stress makes birds more susceptible to disease. The Slatey and Plum-heads are the species in this group susceptible to stress. Stress can come in the form of traffic noise, from dogs, cats or from other predators such as rats and mice at night, from an incorrectly designed aviary, from unhygienic aviary management, from the incorrect diet during the breeding season and from anything which affects the physical and psychological well-being of the bird. Stress control is an important part of disease prevention.

The basic rules used to prevent disease entering an aviary

(a) Purchase healthy and alert birds only.

(b) Do not purchase birds in poor feather or with tail feathers missing, with watery eyes, or going light.

(c) Quarantine all newly purchased birds for six weeks in an isolation cage.
(d) Observe for illness whilst in quarantine.
(e) Ill birds in quarantine must be diagnosed and treated appropriately before entering the aviary.
(f) Only healthy quarantined birds are to enter your aviary.

4. Recognition of disease

Early recognition of disease is extremely important if these diseases are to be controlled. Because birds have the great ability to hide their illnesses, any bird that appears sick must be helped immediately.

Good aviary management must include a proper surveillance of the flock for signs of illness. Signs of an unwell bird are sometimes subtle, however with careful observation they will become very clear and alert you to take immediate first aid steps. It is best to observe the birds from a distance before closer observation of the flock. This way it is possible to see how the birds are behaving in their undisturbed environment. This way any abnormal behaviour can be more accurately determined.

Recognising a healthy bird allows us to be able to detect a bird which is unwell. So, it is important to observe your birds and get to know their normal behavioural patterns.

The signs of illness vary according to the disease involved, but initially there are some general changes which occur in the behavioural pattern and physical appearance of the ill bird. **Watch for a bird:** less active than normal, fluffed up with both feet on the perch, rubbing its head on the perch, with a slight tail bob, with its head behind its wing, sitting low on the perch, which appears sleepy, not eating with the other birds, which does not fly away when you enter the cage, on the ground.

Or the bird may: sneeze, have wet feathers around the head, have a wet eye, have a shut eye, have a droopy wing, have faeces stuck to the vent or tail feathers, be unable to fly.

Once recognised as being ill the bird should be moved to a specially designed hospital cage which should be away from the aviary. This cage provides an opportunity for the bird to rest in a controlled environment, isolates the bird from the rest of the aviary and allows you to more closely observe the bird and to start any intensive treatment which may be necessary.

5. Emergency first aid in the event of disease

Emergency first aid is necessary in this group of parrots once an illness has been detected. Within this group of parrots the Plum-heads and Slaty-heads have adapted successfully to ground feeding in the wild because they are quick witted and alert. Their needs for water are minimal. These birds are difficult to treat because of this "highly strung" nature, because they drink so little and their small size. In the diseased state this spells disaster. Birds rely on body heat, rest, and hydration to overcome illness. The small size of the Slaty-heads and Plum-heads makes them very susceptible to loss of body heat and they find it hard to rest and get proper nutrition and fluids because of their flighty nature. It is very important to have a routine of emergency first aid procedures, so that the bird can be maintained whilst a diagnosis is made. The most important first aid steps are as follows:

(a) Prepare the hospital cage. It is necessary to place the ill bird under the least amount of stress as possible. The bird should be held as little as possible and then placed into the prepared hospital cage. The cage and feed/water containers should be already disinfected, filled with fresh food and water and a piece of clean newspaper or gladwrap placed on the floor in order to collect a fresh dropping sample for analysis.

(b) Move the bird out of the aviary and into the warmed hospital cage. It is best to catch the Asiatics using a net, so as to minimise any unnecessary stress to the ill bird. Examine the bird carefully before placing it into the hospital cage. Write down the findings of this examination noting any symptoms such as weight loss (going light), pasted vent, watery eyes, colour of the dropping ect. which may help the veterinarian make his diagnosis. Provide a quiet environment etc.

(c) Provide controlled heat at about 38-40C.

(d) Observe that the bird is eating and collect a dropping.

(e) Phone your avian veterinarian for advice he may recommend some immediate therapy while the droppings are analysed. It is important to collect a fresh dropping and wrap it in an air tight manner before any medication is given. It is difficult to identify many diseases once medication has been given.

(f) If the bird is not eating or has not eaten it may be necessary to force feed.

6. How to get the best out of your avian veterinarian

The bird patient is unique in medical science. No other animal masks the signs of illness so well. It is thought that birds try to hide symptoms of disease due to their natural fear of being attacked by members of their own flock, since the flock does not want any member attracting predators. It is this innate ability to hide the signs of illness which makes the job of the avian veterinarian so challenging.

A quick and accurate diagnosis is essential if the bird patient is to be saved. The diagnosis of most bird conditions can be made more quickly now that there are reliable diagnostic procedures available (e.g. blood tests, bacterial culture and special stains used on body fluids). However, an accurate diagnosis can be made difficult, if incorrect first aid treatment is given before the veterinarian is consulted.

To get the best out of your avian veterinarian follow the guidelines below:

The flock-health takes precedence over the health of the individual in the aviary situation and sometimes, sad as it may seem, it becomes necessary to sacrifice the life of one ill bird in order to save the rest of the flock.

(a) First assess your disease situation

If one or at the most two birds fall ill then it is more than likely that there is no serious threat to your aviary. Immediately isolate these ill birds and treat them with appropriate first aid (refer to first aid treatment on next page). Contact your avian veterinarian for further advice. It is important to observe the flock carefully over the following two to five days and if further illnesses occur then it is almost certain you have an infectious problem and ill birds should be submitted as soon as possible to the veterinarian for immediate analysis.

(b) Submit appropriate birds to the avian veterinarian

To expedite the correct diagnosis it is imperative that ill or dead birds taken for analysis have not been treated with any medication. Most tests for bacterial infections will be erroneous if medications are given within 48 hours of testing.

It is best to submit an ill live bird to the veterinarian so that he can observe the symptoms and take tests on the live bird. Some diseases are difficult to detect in the dead bird and it may be necessary to examine body fluids from the live bird for accurate diagnosis.

Place clean newspaper or gladwrap on the floor of the transport cage of the ill bird so that droppings can be collected and tested immediately on arrival at the veterinary hospital.

Autopsy cannot help in making a correct diagnosis if the bird has been dead for more than six hours before it is discovered in the aviary or hospital cage. Birds deteriorate very quickly after death and it is important to preserve them as soon after death as possible.

The dead bird must be cooled if the organs are to be preserved for autopsy. This is best achieved by wetting the bird in detergent (to stop the feathers from insulating the body), wrapping the body in paper and then plastic and refrigerating until the body can be taken to the veterinarian. The body is best stored in the vegetable crisper part of the fridge if you are able to get to the veterinarian with four days and if it is longer than this time then it is best to place it in the freezer section.

Transport the body to the veterinarian in a cooler.

A detailed autopsy is sometimes the only way of diagnosing a disease in the aviary. Only a few conditions can be accurately diagnosed at autopsy by looking at the organs alone and it is for this reason that a detailed autopsy is usually required for the exact diagnosis. A detailed autopsy involves taking tests of the body fluids which will give an immediate recommendation for further first aid treatment. For the exact treatment recommendations the organs must be sent to an avian pathologist for microscopic analysis, with the results back in about 48 hours.

Immediate first aid treatment which will not delay an accurate diagnosis

When the bird patient shows signs of illness the situation must be treated as an emergency. The ill bird's temperature drops rapidly once it stops eating and together with dehydration sends the bird into circulatory failure and kidney shutdown. It is for these reasons that immediate first aid treatment (which will not interfere with the diagnostic testing procedures) should be administered.

(a) Isolate an ill bird from healthy birds.

(b) Provide warmth and quiet. Keep the temperature within the cage 80 to 90F (29-30C). This can be accomplished with a heat lamp placed a few feet from the birds cage or there are special hospital cages available for the aviary bird keeper. In either case check the temperature often by observing the bird. If it is still fluffed up the temperature must be increased and if it is holding the wings out or panting it is obviously too hot and the temperature must be decreased.

(c) Provide food and water. Keep food and water in easy reach.

Move the food and rearrange the perches so that the food is constantly in front of the bird. Without food the ill bird's body temperature drops rapidly and without water dehydration occurs. It is a good idea to learn a force feeding technique utilising a crop needle. Most bird clubs have demonstrations on this technique. A good energy mix for the ill bird is glucose and water (1/4 teaspoon of Glucodin powder to 30 mLs of water) given warm by crop needle or placed in the water dish. This solution must be made fresh and replaced within twelve hours.

(d) Do not give any medication such as antibiotics which may interfere with the diagnostic tests.

7. Prevention and control of disease within the aviary

Good management prevents disease within the aviary. Good management takes into account the system of housing and quarantine, aviary hygiene, breeding preferences, proper nutrition, accurate record keeping and an informed policy against disease. By providing the best physical and psychological environment for captive birds, breeding results are maximised and disease minimised. I will discuss the diseases which most commonly occur within the aviary of the Asiatics and how best to control these diseases.

The diseases of note are worm infestations, Papovavirus, bite wounds (especially toe injuries), bacterial infections of the bowel and Aspergillosis. Quarantine helps in preventing the entry of most disease but it can never be 100% effective.

Worm infestation

What type of worms affect the Asiatic Parrots?

These parrots spend some time on the ground in captivity and are exposed to the worm eggs in the droppings. The most common worm in Asiatics is the roundworm (Ascaridia) and hairworm (Capillaria), although they sometimes can become infested with tapeworms.

What are roundworms?

Roundworms are an off-white colour and look like short, thin spaghetti, but are usually not seen in the faeces except following deworming or on post mortem examination, where they can be seen to fill and distend the intestine.

Those birds most susceptible to roundworm infestation are young birds, birds under stress (of any age or species), and those birds which make it easy for the parasite to complete its life cycle (e.g. Asiatic Parrots which spend time on the ground; birds which are not separated from their droppings i.e. not in suspended wired floor aviaries; and birds exposed to wild bird visitors).

What are the signs of roundworms?

Symptoms of roundworm and hairworm infestation are usually obscure, except in heavy infections. Mildly infected birds appear unthrifty and may not breed. In heavier infections the bird will be deprived of valuable nutrients and will eat more but lose weight. This may not be readily apparent by looking at the bird through the aviary wire and most cases of worm infestations are detected when the bird suddenly becomes ill and fluffed up and obvious. The ill bird may have its head behind its wing or become unsteady on its feet and have a "dropped head" appearance. A bird with these advanced symptoms is in grave danger with or without appropriate treatment, because of the danger done to the obstructed bowel. Hairworms are blood sucking worms and the most common sign of infestation is lethargy due to anemia.

How are roundworms and hairworms diagnosed?

Because roundworms are not usually seen in the droppings, the diagnosis of a mild infestation may prove difficult without a microscopic examination of the dropping. Sometimes a bird will have roundworms but the eggs will not be visible under microscopic inspection of the faeces.

When is the aviary treated?

Treatment of the whole flock is recommended four weeks before and immediately after the breeding season. Wet, humid weather favours an outbreak of roundworms and if there has been previous problems with worms in the aviary repeat dosing after such weather conditions is prudent. Remember to repeat treatment fourteen days after dose to kill any immature worms which may have escaped the initial treatment.

What is the treatment?

If roundworm infestation has been diagnosed all birds must be treated at the same time with an anthelmintic (anti-worm medicine) which is active against the roundworm and safe to the species of bird. The ideal anthelmintic is one which acts not only on the adult worm but also on the eggs and larvae. Any introduced birds must be treated for roundworms before entry into the aviary. This is best done whilst in quarantine. Another source of infection are wild bird visitors which can leave infective droppings in the aviary. It is important to stop such visits by sparrows, doves etc. since roundworms are not species specific. Regular microscopic testing can identify an unknown source of infestation.

Treatment may effect a clinical cure but ultimate control rests with management. Safe and effective anthelmintics are available. Administration by drinking water gives disappointing results, therefore gavages (medication by crop tube) will be more effective. Effective anthelmintics are Fenbendazole (25-50mg/kg): Levamisole (10-20mg/kg): Mebendazole (25mg/kg twice daily for five days): and Ivermectin (0.200mg/kg).

Fenbendazole (Panacur 2.5)

Fenbendazole is a very safe drug, however may cause feather problems if given during moult. Oral dosing gives best results at a dose of 0.2mL/100gm in birds less than 400gm and 0.1mL/100gm in birds heavier than 400gm. Given in the drinking water at 5mL/litre changed daily for five days. Panacur will not kill hairworms.

Lavamisole (Nilverm 1.4%)

Levamisole acts on both the immature and mature roundworm as well as the hairworm. The safety margin is quite low so the dose should be measured out carefully. If a serious infestation with roundworms is suspected, Levamisole is not the drug of choice, since all roundworms are killed too quickly and can cause an obstruction. Other anthelmintics do not have this problem. Oral dosage 0.1mL/100gm repeated in two weeks and water dosing is 50mL/litre as only drinking water for 24 hours.

Ivermectin (Avermec)

Has a fairly good margin of safety (30 x), however, care must be taken in diluting the concentrate 100 times with propylene glycol. Ivermectin and propylene glycol have bitter taste so must be given by crop tube. The dose of diluted mix given is 0.2mL/ 100gm body weight repeated in 21 days.

How are roundworms and hairworms controlled?

Whenever a worm infestation occurs, the management must be reassessed and tightened. Control of recurrent problems can be reduced by modern aviary construction with birds housed off the ground or on readily cleaned concrete floors. Concrete floors are easy to clean and disinfect, but may hold moisture at times. Eggs must undergo development in the environment before becoming infective to the next host. Warm and wet conditions enhance this development. Therefore regular scraping and a dry aviary are of utmost importance to prevent the spread within the aviary. Clean aviaries mean healthier birds. Also overcrowding may effect the health of birds. In an aviary with a dirt floor, if a roundworm problem is diagnosed the only way of removing the eggs is by removing the top 10cm (4in) of soil. Never dig up, since this will protect the eggs.

Disinfectants will not kill the eggs in soil. Caustic soda will kill eggs and all vegetation as well.

Rules for the control of roundworms and hairworms in Asiatics.

(a) Control entry of worms. Treat all birds before entering aviary. Prevent wild bird contamination of aviary.

(b) Treat birds in aviary with an appropriate anthelmintic and repeat in fourteen days.

(c) Prevent the worm eggs from larvating. By maintaining dry conditions, removing droppings within 48 hours, or by using elevated wire floor cages. With dirt floor aviaries regular removal of the 10cm (4in) of soil or treatment with soda lime especially after wet weather may be necessary.

(d) Prevent the stress of captivity.

So to lessen the effects of worm infestation. Ensure all birds are on a good balanced diet especially the young birds. Do not overcrowd and prevent inter-species aggression.

Worming Dose Rates

Species	AV. Weight	Name and Volume Of Anthelminthic		
		Levamisole 1.4 (0.1mL/ 100gm)	Panacur 2.5 (0.2mL/ 100gm)	Ivermectin (1:100) (0.2mL 100g)
Alexandrine	230-240g	0.25mL	0.5mL	0.5mL
Derbyan	265-270g	0.3mL	0.5mL	0.5mL
Moustache	125-130g	0.15mL	0.25mL	0.25mL
Ringneck	130-140g	0.15mL	0.3mL	0.3mL
Malabar	80-90g	0.1mL	0.2mL	0.2mL
Plum-heads	70-80g	0.08mL	0.15mL	0.15mL
Slatey-heads	90-120g	0.1mL	0.2mL	0.2mL

All above dose rates are given orally with a crop needle.

Papovavirus
What is Papovavirus?

Papovavirus is a breeding disease seen mostly in nestlings and fledglings.

What are the signs of Papovavirus?

Papovavirus may appear in several forms. Usually there will be sudden death in nestlings. In older birds there may be feather disease similar to PFBDS. A stunted nestling with bloated abdomen and signs of haemorrhage would indicate Papovavirus in the Asiatic Parrot.

What is the treatment for Papovavirus?

There is no effective drug against Papovavirus. If there is a suspicion of Papovavirus an avian veterinarian is necessary to confirm the disease and to advise you on a control and preventative programme.

Bite Wounds

Transport the Asiatics as a single bird to a cage to prevent toe biting in transit. As well keep only compatible pairs together and separate cages with double wire.

Bacterial infections of the bowel

The Asiatics do have problems with bacterial bowel infections. The first signs of illness is diarrhoea followed by lethargy. This is mostly seen in birds fed sprouted seed or appears often in a bird with worms. The treatment is to stop feeding sprouts and a short course of antibiotics.

Aspergillosis

The Alexandrines and Ringnecks appear to be prone to acute infections with this mould. The mould is usually associated with fruit and vegetables left in the aviary for longer than twelve hours. Wet areas and condensation in the aviary will also be predisposed to Aspergillosis. Think of Aspergillosis first in any sudden onset of severe respiratory distress (gaping and mouth breathing) in

Asiatics. A cure is rarely successful without surgery, therefore early veterinary attention will help these birds.

Chlamydiosis

What is Chlamydiosis?

Chlamydiosis is also known as Psittacosis, Parrot Fever and Ornithosis. It is caused by a small, toxic virus - like organism which is capable of causing death and illness in birds and man.

Which species are most commonly affected?

This is a disease seen in neglected Ringnecks and Plum-heads. The birds may have been housed in a large aviary with other species of birds. The birds often have other diseases at the same time but the cause of the illness is usually due to the highly toxic Chlamydial organism.

What are the signs of Chlamydiosis?

The clinical signs of affected birds can vary considerably, however the most consistent sign seen is conjunctivitis. Conjunctivitis may first be noticed by a wet eye, rubbing of the face on the perch then thickening of the eye lid of one or both eyes with loss of feathers around the eyes. Other signs of Chlamydiosis include prominent keel bone, fluffed feathers and a soiled vent. In the most severe form the bird may be found dead on the aviary floor without any previous signs of illness and in good condition.

How is Chlamydios is prevented from entering your aviary?

There are two rules for the prevention of Chlamydiosis:-

(a) Only purchase birds with no signs of Chlamydiosis.

(b) Quarantine all newly purchased birds for six weeks.

When purchasing a new bird check carefully for any signs of eye-watering, feather loss around the eye, soiling of the vent and going light. If the bird shows any of the above signs then it should not be purchased. Unfortunately the disease may still enter your aviary in birds free from the above symptoms, because sometimes the bird will be a "carrier" of the disease. A "carrier" bird is one which appears healthy but is capable of infecting other birds. A "carrier" bird may become ill when stressed. Changing aviaries is stressful to birds and often a healthy "carrier" will become ill when introduced to a new aviary.

Chlamydiosis can be prevented from entering your aviary by placing all new birds in quarantine for six weeks. Within this time most birds carrying Chlamydiosis will fall ill. If this occurs a long course of chlortetracycline will rid the bird of the disease.

Psittacine "Feather and Beak" Disease (PFBDS)

What is Psittacine "Feather and Beak" Disease?

Feather and beak disease is most well known from the syndrome seen in the Sulphur-crested Cockatoo. It is a virus infection affecting the feathers and beak in the cockatoo, however in the Asiatics the same virus causes a feather and nestling disease with no beak involvement.

Which species are affected?

The disease is seen in all of this group of parrots, and

especially in Asiatics in mixed collections housed together with infected Cockatoos, Princess Parrots and/or Neophemas.

Slatey-heads with this infection will have poor tail feathers and look untidy.

What are the signs of PFBDS?

In this group of birds the most outstanding sign of the disease is "shabby" plumage and feather loss around the head. It is usually a young bird disease seen most often during the first juvenile moult.

How is PFBDS prevented from entering the aviary?

Birds known to be in contact with "runners" or birds with signs of PFBDS, birds in poor feather or without tail feathers should not be purchased. Quarantine does not help prevent this disease as it will not show up until the next moult or during the breeding season. "Carrier" birds do occur so it is important to buy your birds from an aviary known to be free of this disease.

Is there a treatment for PFBDS?

Because the disease is caused by a virus there is no known treatment at the present time. A vaccine to protect from this disease is not available.

"New Wire" disease

The Asiatics are inquisitive and like to nibble in a new aviary, which predisposes these birds to "new wire" disease. "New wire" disease is a heavy metal (Zinc) poisoning caused by ingestion of the zinc coating on new wire or ingestion of fresh solder joining the wire. It is advisable not to house a newly purchased Asiatic in a cage or aviary which has not been weathered or "deactivated" with a wire scrub of vinegar.

Signs of "new wire" disease will appear within a few days to weeks depending on the amount of "new wire" ingested. The first sign may be "sleepy" eyes in the bird but may quickly develop into vomiting, lethargy followed by paralysis and death.

Veterinary treatment will save most birds suffering from "new wire" disease, as long as the kidneys have not been irreversibly damaged. Early veterinary attention, therefore, will preserve your investment.

Informed disease control

Whenever a bird becomes ill or dies it is important that the disease is accurately identified, so that the disease status of your aviary is known. This allows your avian veterinarian to formulate a disease control programme to protect the rest of the aviary at this time and in the future. Good record keeping allows the disease process to be understood quickly so that the spread of the disease can be halted rapidly.

To prevent a disease from spreading through the aviary surveillance for ill birds should be part of a good management programme. The ill bird should be isolated into a warmed hospital cage and the disease identified by an avian veterinarian before any medication is prescribed. Indiscriminate use of drugs is dangerous as it makes diagnosis of the disease difficult and may lead to resistance of the disease organisms to certain antibiotics.

Any dead birds should be autopsied to determine the cause of death so that appropriate treatment can be started.

Rules on the prevention of disease within the aviary
(a) Good Management
(b) Proper Nutrition
(c) Informed Disease Control

8. Best methods of dosing and administration of drugs

The most useful method of dosing the Asiatic parrots is by crop tube (size 14). The advantages over water medication are that the exact dose can be given quickly and with a minimal of stress to the bird. The disadvantages are that the birds must be caught and medicated individually, which is time consuming and at times stressful to the birds (especially during the breeding season). However, the Asiatic Parrots drink very little which makes crop tube medication necessary.

The method of using the crop needle.

It is best for someone to demonstrate this technique. The basic procedure is to place the crop needle into either side of the mouth and to pass high to the roof and towards the left of midline of the back of the mouth. Follow the food pipe (oesophagus) down and gently move into the crop. Once in the crop the end of the needle can be felt through the crop wall. If it is in the trachea you will feel no needle tip and the trachea may be bent. Once into the crop the medication or force feeding mixture is slowly delivered and the crop needle removed. Reflux occurs if the medication is delivered too quickly. The mixture will ooze from the mouth. If this happens quickly remove the needle and point the bird's head towards the ground so that the mixture falls out and does not enter the lungs.

The size of the crop dictates the volume of mixture or medication that can be given. The adult bird has a much smaller capacity than the nestling. The crop capacity of the adult Plum-head is about 2mL and the adult Alexandrine is about 8mLs, whereas a baby Ringneck can take 10-20mLs of food. The maximum volume to be instilled into the crop of the adult Slatey-head is 2mLs, the nestling Ringneck 15mLs, the adult Alexandrine is 5mLs. If too much volume is given the medication may enter the lungs and cause pneumonia and death.

The dangers with this method disappear with practice and proper tuition. All clubs should demonstrate this technique to their members. A good practice is to use sterile water until your technique is polished. Sterile water will not cause pneumonia if only a small amount enters the lung.

Proper use of drugs

Nowadays we are lucky to have good drugs available and most bird diseases can be treated successfully. Know the weight of the bird and dose accordingly. Use an accurate measure e.g. calibrated syringe and crop needle.

9. Diseases of baby Asiatics

Lameness and inability to fly (Rickets)

The most common disease seen in parent reared Asiatics is Rickets, which is caused either by inadequate Calcium in the diet or a bacterial infection of the bowel. The bowel infection may well be originating in the sprouted seed mix. The symptoms of rickets include abnormal legs (splayed), rubbery beaks, lameness and the inability to fly. This disease can be prevented by giving additional Calcium in the water during the feeding of the young or by Calcium injections to those birds which are badly affected. It is necessary to consult a veterinarian, if there is an incomplete response to Calcium supplementation.

Vomiting (Food too Cold, Thrush, Megabacteria, Papovavirus)

The feeding formula must be fed to the young bird hot (40C). Crop disease occurs when the food is fed too cold which causes vomiting. As well, hand-reared babies often develop vomiting due to thrush. Thrush is usually related to improper hygiene with the feeding utensils. Treatment is successful if there are no other diseases present. Use Nystatin for thrush according to the directions on the prescription. If there is no response to the Nystatin then Megabacteria or Papovavirus may be the cause of the vomiting and you should consult an avian veterinarian immediately.

Any hand-reared baby which starts to vomit must be isolated from the other babies and fed last, until the cause of the vomiting is known.

Stunting and poorly developed feathers

These symptoms may be related to inadequate nutrition (feeding formula is too watery), worms, bacterial infections or Papovavirus. Any hand-reared baby with these symptoms must be isolated and fed alone and after the other babies. Your avian veterinarian should be consulted. This is seen especially in the Plum-heads and Slatey-heads.

Going Light

This is common in fledglings and may be a normal adjustment to the weaning process. As well, it may be worms or bacterial infections. Your avian veterinarian should be consulted.

10. Surgical Sexing

This is a safe procedure performed under a gas anaesthetic which allows birds to be sexed at a young age, giving a breeder the opportunity to select definite pairs before the breeding season. This is especially important with this expensive group of birds. It requires an veterinarian to perform the procedure. A very small fiber optic light (endoscope) is inserted into the abdominal cavity through a small incision and the sex organs are visualised.

The Asiatics take from two to three years to reach sexual maturity and the sexing procedure is most reliable in birds older than six months of age. The procedure is sometimes inconclusive in birds younger than 6 months, because the sex organs have not become large enough to identify positively.

During the procedure other organs are examined for

abnormalities and the bird's breeding capability can also be determined.

Additional Basic Notes on Bird Diets

A seed diet alone is insufficient for birds. During breeding and moulting the protein level required by the birds is not met. Many of the essential amino acids that make up proteins are low or non existent. Vitamin content is variable and the mineral content depends on the ground where the seed was grown. The basic seed diet should therefore be supplemented. The authors find milk thistle and oats in head (grown on home ground), corn on the cob and some fruit and dry dog food meets their birds' requirements. Some other foodstuffs beneficial to your birds are listed below. It should be noted that there are a great many foods that will be eaten by your birds but they need only be fed a few of these items providing they ensure the balanced diet of vitamins, minerals and proteins the birds require.

Some Foods beneficial to your birds:

For vitamins A, D3 and E in order of richness in Vitamin A.

Carrots (best boiled for a few minutes to release Vitamin A)
Broccoli
Silverbeet (Silverbeet contains oxalic acid which prevents the absorption of calcium by the bird. To avoid this problem Silverbeet should be boiled for a few minutes, certainly before being fed to laying hens or young growing birds).
Paw Paw
Watermelon

Vitamin B

Seeding grasses
Sprouted seeds
Dark green vegetables

Vitamin E

Wheat Germ

More protein is necessary for birds that are moulting or breeding. So once a week give one hard boiled egg yolk and a cooked chicken leg from the dinner table between a pair of birds. Egg yolk provides animal protein and Vitamins A, D, E, K and B. Chicken bone provides animal protein and minerals. When the birds are moulting Niga and Rape Seed should form about 2% of the seed mix.

Handraising Asiatics

This section is a simple guide to aid the reader who may find him or herself in the position of foster parent. The method put forward here is adequate for baby birds seven days old or older. Rearing babies from the egg or one or two days old is a lot more complicated to say the least. Eggs or very young birds should be fostered out to the nests of similar species or to Fischers or Peach-faced Lovebirds until at least a week old whenever possible.

When the baby bird or birds are removed from the nest they

should be taken inside and placed in a warm receptacle. Some people have themostatically controlled brooders. Most people do not. This wonderful appliance is fortunately not necessary for Asiatics a week or more of age especially if there are two or three or more babies that can huddle together to keep warm. All that is required is a suitable container and a heat source. A cardboard box is quite sufficient. It holds heat and moisture, the lid can be closed to keep the babies in the dark as in their natural nest, boxes can be burned when dirty and last but not least they are readily available in all sorts of sizes.

Mention the words heat source and many people will think - "light bulb". Light bulbs do give off heat but they also shed light which has to be unpleasant to the babies. To find a way to get heat but no illumination from a light bulb is quite a task and can lead to a smoke-filled room or worse. A couple of much simpler methods are (a) sit the box or boxes on an electric blanket (covered with a sheet or a square of cardboard or the like). This can be kept on low heat - especially at night. (b) Take the element out of a warming slipper, wrap it up in newspaper and tape. Put it in the box leaning against one side. The birds will sit beside it or move away from it as they wish. It will not catch fire. Syd and Jack Smith have been using this method very successfully. Depending on where you live or the time of year artificial heat may not be required at all.

Now that the babies are comfortably housed leave them alone for a few hours. Having just moved to a new environment babies are more inclined to eat well for the first time when they are really hungry. Obviously if the crops are empty and the babies are starving they should be offered food straight away.

Preparing the rearing mix.

The following mixture is as simple as possible. The ingredients are readily and conveniently available.

Ingredients:-

2 parts Heinz high protein baby cereal
1 part Uncle Toby's rolled oats
1 part Sunflower Kernels
Powdered Cuttlebone
Pentavite multi-vitamin liquid
Silverbeet water

Method

Dry Mixture

Place the required amount of Sunflower Kernels in the blender and blend. Add the rolled oats and blend. If there is room add the high protein baby cereal and cuttlefish powder at the rate of 2 tsp per 500g of mix.

The baby cereal and cuttlefish powder doesn't need blending but the blender makes a good and easy job of thoroughly mixing the ingredients together. Place the mixture in an air-tight container until required.

Silverbeet water

Boil a Silverbeet leaf for a few minutes. Place it in the blender with a little water and blend. Finely sieve the liquid into a jug or cup.

Feeding Time

First of all put the kettle on.

Put the desired amount of dry mixture, trial and error initially, in a cup or small bowl. Add some cold silverbeet water and stir into a thick paste. Cold water prevents lumps forming. Add boiling water from the kettle to desired consistency (so it will slowly pour off a teaspoon like yoghurt).

Place the cup in a bowl of boiling water to keep it hot when feeding the birds. Add 2 drops of Pentavite and stir in one meal per day. The best utensil for feeding birds such as these is a 'bent' teaspoon. Place a spoon on its side in a vice or similar and tighten down until the sides of the spoon bend in.

Feeding the birds

Lift one bird out of the box place it on a towel or sponge so it can't slide about. Place one hand over the birds back supporting its head - if required - between thumb and forefinger. Test the temperature of the food against your top lip. It should be very warm but not scalding. Put the end of the spoon under the tip of the baby's top mandible and gently raise the spoon and pour the food into its mouth. Babies take to the spoon very quickly and feeding will soon present no problems, quite the opposite in fact. This is why the birds should be fed one at a time. Three babies fighting over a food laden spoon becomes very messy. If the baby refuses to eat, nine times out of ten it is because the food is not warm enough. After feeding wipe any excess food from the baby's face or throat. If the food is anywhere else on the bird get your wife to take over feeding. The crops should be seen to be full, not bloated after a feed. The babies should be fed when their crops are 90% empty. This will be roughly 3-5 feeds per 24 hours depending on their age. The first feed will be when one gets up in the morning, the last feed just before one goes to bed. As the birds get older the food mixture can be made thicker and feeds will be less frequent. As they feather up and become more recognisable as birds they will not require heat unless it is really cold. They should be placed in a wire cage or the like. A cage without a base sitting on sheets of newspaper is very convenient for cleaning etc. Canary seed, sunflower kernels, sprouted mung beans some egg and biscuit etc can be placed on the paper or in shallow bowls or trays. The birds will start pecking at their food within a few days. They will start to refuse spoon feeds and in the mornings one will feel hard seeds in their crops. There is really no problem in weaning birds. The birds will do it all by themselves.

Publishers Note

This title is published by Australian Birdkeeper Pty Ltd who produce a wide and varied range of avicultural literature including the world acclaimed Australian Birdkeeper magazine - a full colour, 48 page bi-monthly magazine specifically designed for birdlovers and aviculturists. It is our intention to produce high quality, informative literature for bird lovers, fanciers and aviculturists alike throughout the world. It is also our belief that the dissemination of qualified information on the care, keeping and breeding of birds is imperative for the total well-being of captive birds and the increased knowledge of aviculturists.

We wish you all the best and Happy Birdkeeping.

Nigel Steele-Boyce
Editor-In-Chief
Australian Birdkeeper